Sediment Transport - Recent Advances

Edited by Andrew J. Manning

Published in London, United Kingdom

IntechOpen

Supporting open minds since 2005

Sediment Transport - Recent Advances
http://dx.doi.org/10.5772/intechopen.91593
Edited by Andrew J. Manning

Contributors
Mimouna Anhichem, Samir Benbrahim, Georgii V. Nesyn, Bernhard Vowinckel, Tian-Jian Hsu, Eckart
Meiburg, Leiping Ye, Kunpeng Zhao, Bofeng Bai, Andrew J. Manning, Shu-Qing Yang, Claire Chassagne,
Zeinab Safar, Zhirui Deng, Qing He, Ahmad Shakeel, Alex Kirichek, Katherine Cronin, Thijs van Kessel,
Lynyrd De Wit, Deyan Draganov, Xu Ma, Menno Buisman, Tjeerd Kiers, Karel Heller, Ishraq AL-Fadhly

Notice
Statements and opinions expressed in the chapters are these of the individual contributors and not
necessarily those of the editors or publisher. No responsibility is accepted for the accuracy of
information contained in the published chapters. The publisher assumes no responsibility for any
damage or injury to persons or property arising out of the use of any materials, instructions, methods
or ideas contained in the book.

First published in London, United Kingdom, 2022 by IntechOpen
IntechOpen is the global imprint of INTECHOPEN LIMITED, registered in England and Wales,
registration number: 11086078, 5 Princes Gate Court, London, SW7 2QJ, United Kingdom
Printed in Croatia

British Library Cataloguing-in-Publication Data
A catalogue record for this book is available from the British Library

Additional hard and PDF copies can be obtained from orders@intechopen.com

Sediment Transport - Recent Advances
Edited by Andrew J. Manning
p. cm.
Print ISBN 978-1-83881-118-1
Online ISBN 978-1-83881-119-8
eBook (PDF) ISBN 978-1-83881-120-4

We are IntechOpen,
the world's leading publisher of
Open Access books
Built by scientists, for scientists

6,000+
Open access books available

148,000+
International authors and editors

185M+
Downloads

Our authors are among the

156
Countries delivered to

Top 1%
most cited scientists

12.2%
Contributors from top 500 universities

Interested in publishing with us?
Contact book.department@intechopen.com

Numbers displayed above are based on latest data collected.
For more information visit www.intechopen.com

Meet the editor

Professor Andrew J. Manning is a principal scientist (rank grade 9) in the Coasts and Oceans group at HR Wallingford, UK. He has more than 23 years of scientific research experience in both industry and academia examining natural turbulent flow dynamics, and fine-grained sediment transport processes, and assessing how these interact using both field studies and controlled laboratory flume simulations. He also lectures in coastal and shelf physical oceanography at the University of Plymouth, UK. Internationally, Professor Manning has been appointed visiting/adjunct/guest professor at the University of Hull, UK; Stanford University, USA; University of Delaware, USA; University of Florida, USA; and Delft University of Technology, Netherlands. He is a highly published and world-renowned scientist in the field of depositional sedimentary flocculation processes. He is a fellow of the Royal Geographical Society and recipient of a University of Plymouth Vice Chancellor's Research Fellowship (2007) and the "Exemplary Act Award" by the US Department of the Interior and US Geological Survey (2015). He has contributed to more than 100 peer-reviewed publications in marine science, more than 50 of which have been published in international scientific journals, in addition to more than 140 articles in refereed international conference proceedings. He supervises graduates, postgraduates, and doctoral students focusing on a range of research topics in marine science. Professor Manning has led numerous research projects investigating sediment transport and dynamics in aquatic environments around the world, including estuaries, tidal lagoons, river deltas, salt marshes, intertidal, coastal waters, and shelf seas.

Contents

Preface

The effective management of many aquatic environments requires a detailed understanding of sediment transport and dynamics. This has both environmental and economic implications, especially where there is any anthropogenic involvement. Numerical models are often the tool used for predicting the transport and fate of sediment movement in these situations, as they can estimate the various spatial and temporal fluxes. However, the physical sedimentary processes can vary quite considerably depending upon whether the local sediments are fully cohesive, non-cohesive, or a mixture of both types. For this reason, for more than half a century, scientists, engineers, hydrologists, and mathematicians have been conducting research into the many aspects that influence sediment transport. These issues range from processes such as scour, erosion, and deposition, to how sediment process observations can be applied in sediment transport modeling frameworks. Written by experts in the field, *Sediment Transport - Recent Advances* draws on international scientific research to examine the following sediment transport-related issues: mud rheology, port and waterways maintenance, steady and unsteady flow, fluid mud monitoring, flocculation processes, sediment, and water quality.

This book includes nine chapters written by an international group of research scientists who specialize in sediment dynamics, geomorphology, water quality, rheology, and numerical modeling. Most of the chapters are concerned with sediment transport-related issues in estuarial, coastal, or freshwater environments. For example, there is a chapter on mud rheology in ports and waterways and a chapter on sediment quality in the Bay of Dakhla, Morocco. Other chapters in the book discuss non-intrusive seismic monitoring of fluid mud, sediment removal from oil storage tanks, and formulae of sediment transport in both unsteady and steady flows.

This book is an excellent source of information on recent research on sediment transport. I would like to thank all the authors for their contributions, and I highly recommend this textbook to both scientists and engineers who deal with related issues.

Andrew J. Manning
HR Wallingford Ltd,
Coasts and Oceans Group,
UK

University of Hull,
UK

University of Delaware,
USA

University of Florida,
USA

Stanford University,
USA

Technical University Delft,
Netherlands

University of Plymouth,
UK

Chapter 1

Formulae of Sediment Transport in Steady Flows (Part 1)

Shu-Qing Yang and Ishraq AL-Fadhly

Abstract

This paper makes an attempt to answer why the observed critical shear stress for incipient sediment motion sometimes deviates from the Shields curve largely, and the influence of vertical velocity is analyzed as one of the reasons. The data with $d_{50} = 0.016 \sim 29.1$ mm from natural streams and laboratory channels were analyzed. These measured data do not always agree with the Shields diagram's prediction. The reasons responsible for the deviation have been re-examined and it is found that, among many factors, the vertical motion of sediment particles plays a leading role for the invalidity of Shield's prediction. The positive/negative deviations are associated with the up/downward vertical velocity in decelerating/accelerating flows, and the Shields diagram is valid only when flow is uniform. A new theory for critical shear stress has been developed, a unified critical Shields stress for sediment transport has been established, which is valid to predict the critical shear stress of sediment with/without vertical motion.

Keywords: critical shear stress, non-uniform flows, shields diagram, vertical velocity, decelerating/accelerating flows

1. Introduction

The incipient motion of sediment is one of the most important topics in sediment transport. Generally, two methods are available in the literature to express quantitatively it, the shear stress approach and velocity approach [1]. The latter assumes that if the mean velocity excesses its critical velocity, then the sediment motion can be observed. The former used by researchers represents the force acting on a particle. Shields [2] is the earliest one who used the shear stress approach, or Shields number $\tau/(\rho_s-\rho)gd_{50}$ versus the Reynolds number, and he obtained a famous Shields curve to express sediment initiation. Francalanci et al. [3] interpreted the Shields number as the ratio of streamwise/vertical forces using the following form:

$$\tau_{*c} = \frac{\tau_c}{(\rho_s - \rho)gd_{50}} = \frac{4}{3} \frac{\tau_c \frac{\pi}{2}\left(\frac{d_{50}}{2}\right)^2}{(\rho_s - \rho)g\frac{4\pi}{3}\left(\frac{d_{50}}{2}\right)^3} \tag{1}$$

where τ_c ($=\rho u_{*c}^2$) is the critical shear stress for the median grain size of sediment, d_{50}; g is the gravitational acceleration; u_{*c} is critical shear velocity; ρ_s and ρ are the sediment and fluid densities, respectively. The shear stress exerted by the fluid must be higher than the critical shear stress τ_c to initiate sediment motion at the bed. Based on available experimental data, Shields in 1936 found that the Shields number τ_* depends on the particle Reynolds number (R_*), i.e.,

$$R_* = \frac{u_{*c}d_{50}}{\nu} \tag{2}$$

where ν is the kinematic viscosity of the fluid.

The original Shields diagram has been reproduced and modified by many researchers. A comprehensive review has been done by many researchers [4, 5], in which some significant deviations of the observed critical shear stress from the standard Shields curve were observed. This has attracted extensive research by notable investigators, and some factors leading to the data scatter have been identified and discussed.

Some researchers believe that the definition of the incipient motion may cause the invalidity of Shields diagram, as the incipient motion depends more or less on the experimental observers' subjective judgment. To address this, criteria like "individual initial motion", "several grains moving" and "weak movement" has been introduced to express the incipient motion [6]. Subsequently, an error band has been included in the modified Shields diagram [7].

Other researchers attribute the large discrepancy to the stochastic nature of turbulence and sediment shape, its orientation, or exposure, protrusion [8–11]. It is natural to expect that when sediment is non-uniform, the critical condition is very difficult to determine, as the larger particles could move relatively easily than the finer one that is sheltered [12].

Over the past eight decades, the incipient motion has been extensively studied again and again [4], because the Shields diagram has been found invalid to predict the critical shear stress of sediment transport in some circumstances. The invalidity is not fully explained, some researchers ascribe it to sediment's characteristics, the other believe these deviations are caused by the flow conditions i.e., non-uniformity of flow [13].

Iwagaki [14] firstly linked the wide scatter in Shields diagram with flow's non-uniformity based on his observation: for the same sediment by the same experimenter, the observed critical shear stress in non-uniform flows largely deviates from that in uniform flows. Afzalimhr et al. [13] confirmed Iwagaki's results, they found experimentally that in decelerating flows, the critical shear stress is considerably below the Shields' prediction, and their experimental data are in complete disagreement with the Shields diagram. Other experimental researchers [15, 16] obtain similar results as Afzalimhr et al.'s [13] who claimed " ... there is no universal value for τ_* ". Likewise, Buffington and Montgomery [4] also agreed "less emphasis should be given on choosing a universal τ_* ".

Some researchers try to explain the large discrepancy between predicted and measured critical shear stress by considering channel's characteristics, such as the channel shape and channel slope [16–22]. "the well-known Shields criterion is insufficient for large slope" was observed by Graf and Suszka [23], while Lamb et al. [24] comprehensively re-visited and examined almost all published datasets, and concluded that the critical shear increases with channel slope, this is totally different from the common sense that predicts increased mobility with increasing channel slope due to the added gravitational force in the downstream direction. But Chiew and Parker's experiments [17] in very steep channels show that the critical shear stress is decreased, contrary to Lamb et al.'s [24] conclusion.

Therefore, the brief literature review shows that Shields diagram cannot predict the critical shear stress well and there are many different potential causes for the deviation, among them, it is necessary to clarify how the channel-bed slope and non-uniformity of flow affect the critical shear stress for sediment motion. The primary objectives of the present study are to

1. investigate the mechanism that causes the invalidity of the Shields curve for the incipient motion of sediment transport;

2. examine why the Shields number depends on the water depth's variation or channel slope;

3. establish a universal Shields diagram that is valid for all data available in the literature; and

4. verify the newly established equations using data from the literature.

2. Theoretical considerations of influence of vertical velocity on the critical shear stress

The author has been systematically investigating the role of vertical velocity on the mass and momentum transfer and has obtained a series of important and interesting conclusions [25–27]. It is found that omission of vertical velocity in our existing theorem of sediment transport makes many phenomena unexplainable. For example the presence of vertical velocity in non-uniform flows leads to the deviation of measured Reynolds shear stress from the linear distribution from the free surface to the bottom, consequently the upward velocity causes the positive deviation of velocity from the log-law or the wake-law is needed to express the velocity distribution, and the downward velocity results in the dip-phenomenon, or the maximum velocity is submerged and does not occur at the free surface as the log-law predicts. As the momentum and mass transfers are closely related to each other, it is interesting to investigate how the vertical motion affects sediment transport.

As a continuous effort, this study investigates the influence of upward/downward velocity on sediment incipient motion and the validity of Shields' diagram. **Figure 1** shows how a river flow interchanges with groundwater and the Darcy law

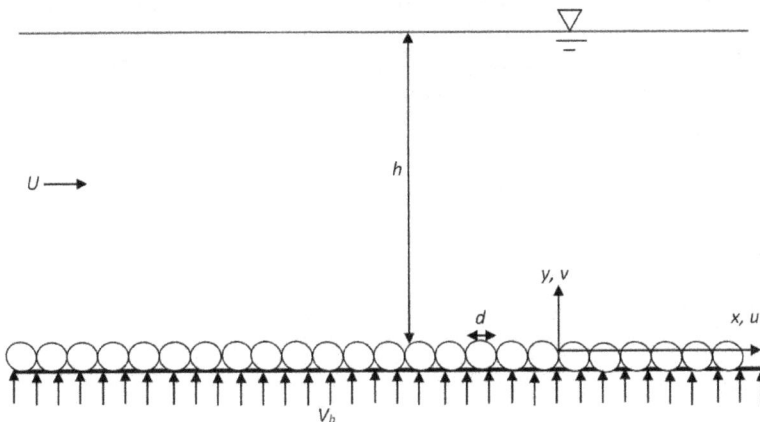

Figure 1.
The upward and downward vertical velocity generating from seepage face injection seepage.

tells that vertical velocity is proportional to the hydraulic gradient, i.e., the suctions and injections inside groundwater can be expected in flood/dry seasons alternatively. The upward flow or injection flow may increase the sediment particles' mobility, or the required critical shear stress is reduced due to the "buoyant effect", which reduces the net settling velocity, mathematically

$$\omega' = \omega - V_b \tag{3}$$

where ω = particle's settling velocity in still water and ω' = the net settling velocity subject to the vertical velocity of groundwater, V_b. The submerged weight in Eq. 1 can be represented by a drag force with the falling velocity ω in still water ($V_b = 0$) as:

$$C_d \pi \frac{d^2}{4} \frac{\rho \omega^2}{2} = \pi \frac{4}{3} \left(\frac{d}{2}\right)^3 g(\rho_s - \rho) \tag{4}$$

where d is the particle diameter, C_d is the drag coefficient.

If the upward velocity V_b of groundwater is so high and $V_b = \omega$, the net settling velocity of the particle becomes zero, thus the particle is neutrally suspended, i.e., liquefaction state. This is often observed during earthquakes. In such case, saturated soil loses its strength and stiffness, it is natural that the Shields diagram cannot predict the particle's critical shear stress. Similarly, if the groundwater in **Figure 2** is downward, then the net falling velocity ω' should be higher than ω, the threshold critical shear stress should be unpredictable using the existing Shields diagram.

The above discussion clearly demonstrates that velocity V_b in a sediment layer may cause the invalidity of Shields diagram, which is supported experimentally by many researchers [28] who conducted experiments by observing the critical shear stress subject to injection and suction flows. Lu et al. [29] has reviewed these experimental results comprehensively. The influence of vertical motion on the critical shear stress has been discussed by many researchers, the parameters used to express the vertical motion include (i) the hydraulic gradient of seepage, e.g., Cheng and Chiew [30]; (ii) the pressure variation in flows [3]. But, there is no research available to investigate the role of time-averaged vertical velocity on the incipient motion of sediment transport.

The introduction of apparent sediment density is similar to Francalanci et al.'s treatment [3]. Instead of modifying the sediment density, they modified the water's density to eliminate the effect of pressure variation over time and space (like pressure induced by waves or bridge piers) on sediment's critical shear stress. Their results show that higher pressure yields higher "apparent water density", and lower pressure corresponds to lower "apparent water density". They found that the

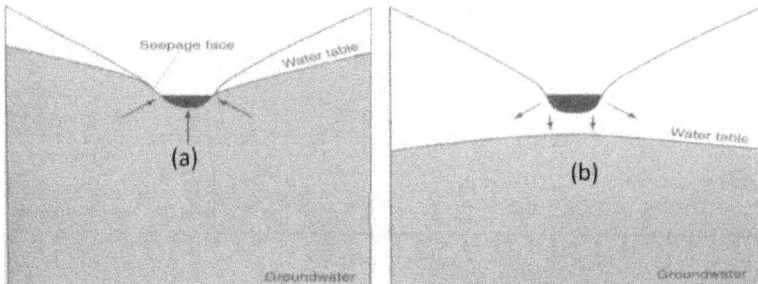

Figure 2.
Schematic diagrams showing interaction of streamwise and vertical motions after Ladson (2008), p99.

Shields number shown in Eq. 1 is actually the ratio of friction force in the streamwise direction (i.e., $\tau_c \pi d^2/2$) to the net force in the vertical direction, i.e., downward gravitational force ($=\rho_s g \pi d^3/6$) minus the upward buoyant force ($=\rho g \pi d^3/6$), so that the effective density of the particle is reduced from ρ_s to $\rho_s-\rho$. When a particle is experienced in the environment with upward velocity V_b, additional upward force will be generated, and the problem is how to determine the additional upward force F_{vb} induced by the upward velocity.

Eq. 4 shows that the denominator of Shields number, i.e., $(\rho_s-\rho)gd$ can be replaced by $3C_d\rho\omega^2/4$, this is why the parameter of settling velocity ω can be used in the study of sediment incipient motion. Some researchers believe that ω is a parameter for suspended load and it should not be used to express the sediment initial motion. The first one who uses ω to discuss the critical velocity is Yang [1], and this treatment significantly simplifies the problem. But this treatment has been blamed by many researchers who believe that the initiation problem does not involve any settling process. Now, Eq. 4 clearly shows that it is logical to express the incipient motion with sediment settling velocity.

For the case shown in **Figure 2**, if the upward velocity is zero, this is a static problem and the net force balance in vertical direction is expressed in Eq. 4. When the upward velocity is non-zero, this becomes a dynamic problem where the lift force F_{vb} should be included, i.e., submerged weight minus F_{vb} must be balanced by the drag force with settling velocity ω'.

$$C_d'\pi\frac{d^2}{4}\frac{\rho\omega'^2}{2} = \pi\frac{d^3}{6}g(\rho_s-\rho) - F_{vb} \tag{5}$$

where $F_{vb} = C_{d,lift}\pi d^2\rho V_b^2/8$. Let $\rho_s' = \rho_s - F_{vb}/(\pi d^3 g/6)$ and inserting the apparent density of sediment into Eq. 5, then the net force in the vertical direction can be alternatively expressed as:

$$C_d'\pi\frac{d^2}{4}\frac{\rho\omega'^2}{2} = \pi\frac{d^3}{6}g(\rho_s'-\rho) \tag{6}$$

In this study, apparent sediment density is introduced, and it depends on the vertical velocity of groundwater. Therefore, it is expected to have a relationship between the apparent sediment density and the settling velocity, similar to Einstein's relativity theory that the length/time depends on velocity if the light's speed is assumed to be constant. Therefore, the effect of vertical motion caused by pressure variation or seepage on sediment transport is eliminated after the introduction of apparent density. In other words, real lightweight particles motion must be the same as those with reduced settling velocity ω' in terms of mobility when both have the same settling velocity. Hence, the apparent density can greatly simplify the mathematical treatment for the complex F_{vb} induced by vertical motions. From Eqs. (4) and (5), the relationship between the modified settling velocity and the apparent density can be expressed by:

$$\frac{\rho_s'-\rho}{\rho_s-\rho} = \alpha\left(\frac{\omega-V_b}{\omega}\right)^2 \tag{7}$$

where α is a coefficient ($= C_d'/C_d$) and $\alpha = 1$ is assumed in this study to simplify the mathematical treatment. Because the drag coefficient depends on Re, this assumption is approximately correct if the particle size is very coarse or the Re does not change significantly with/without V_b. Eq. 7 tells that if V_b is equal to zero, then ρ_s' is the same as the natural sediment; if V_b is positive or upwards then ρ_s' is less than the density of natural sediment ρ_s, and the particles behave like "lightweight sand" or plastic sands; if $V_b = \omega$, then ρ'_s is same as the density of water or similar to neutrally buoyant milk; if V_b is negative or downward, the higher apparent density of sediment behaves like heavy metals.

The vertical velocity V_b in **Figure 2** has the similar effect for the particles' stability as the buoyancy effect, i.e., the submerged weight of the particles is no longer $\rho_s - \rho$, but $\rho_{s'} - \rho$, one may give the general expression of Shields number

$$\tau'_* = \frac{\tau'_c}{(\rho'_s - \rho)gd_{50}} \tag{8}$$

Substituting Eq. 7 into Eq. 8, one obtains:

$$\tau'_* = \frac{\tau'_c}{(\rho_s - \rho)gd_{50}} \left(\frac{\omega}{\omega - V_b}\right)^2 \tag{9}$$

or

$$\tau'_* = \frac{\tau'_c}{(\rho_s - \rho)gd_{50}} \left(\frac{1}{1 - Y}\right)^2 \tag{10}$$

where $Y = V_b/\omega$.

Eq. 9 or 10 generally expresses the influence of vertical velocity V_b on the critical shear stress. It is clear that V_b can be induced by seepage in the sediment layer, it can be also inferred that V_b can be estimated by the Darcy Law using the hydraulic conductivity and hydraulic gradient. Obviously, $Y = 1$ means that the particles can be suspended in water, i.e., liquefaction. If $Y > 1$, it means that particles flow in the upward direction with a net velocity of $V_b - \omega$, this may have a devastating impact on dikes in flood defense as it may cause piping failure. In the following section, the analysis shows that the vertical velocity, V_b is ubiquitous in open channel flows, which is induced by non-uniform flows.

3. Influence of non-uniform flow on the critical shields stress

Ideal uniform flow is very rare in natural conditions, flow rate and water depth/channel width keep always changing, i.e., non-uniform, as shown in **Figure 3**. It is interesting to discuss how accelerating or decelerating flows generate the vertical

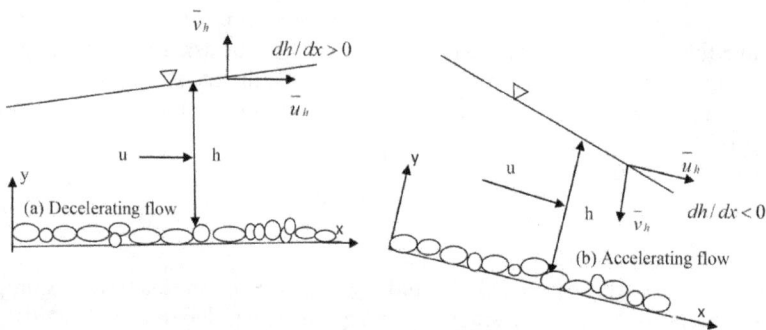

Figure 3.
Non-uniform flows in open channel and the variation of water depth, in which u and v are mean velocities in x and y direction, respectively.

velocity. To simplify the discussion, it is assumed that the flow rate is constant, i.e., dh/dx ($\neq 0$). The 2-D continuity equation is:

$$\frac{\partial \bar{u}}{\partial x} + \frac{\partial \bar{v}}{\partial y} = 0 \tag{11}$$

where $\bar{u}$ and $\bar{v}$ are the mean local velocity at any point in x and y directions, respectively. By integrating Eq. 11, one has

$$\bar{v} = -\int_0^y \frac{\partial \bar{u}}{\partial x} dy \tag{12}$$

$\partial \bar{u}/\partial x > 0$ means accelerating, thus Eq. 12 tells that accelerating flows yield a negative or downward vertical velocity; but decelerating flows generates the positive or upward vertical velocity, i.e., $\partial \bar{u}/\partial x < 0$. Hence, the vertical velocity can be generated by in non-uniform flows.

For a channel with a constant width, its discharge per unit width can be expressed by:

$$Q/b = Uh \tag{13}$$

where Q = discharge; U is the depth-averaged velocity, b is the channel width and h is the water depth. If Q/b could be constant in x direction, one has:

$$\frac{d(Uh)}{dx} = \frac{d}{dx} \int_0^h \bar{u} dy = 0 \tag{14}$$

The vertical velocity $\bar{v}_h$ at the free surface can be obtained from Eq. 12 using Leibniz's rule, i.e.

$$\bar{v}_h = -\int_0^h \frac{\partial \bar{u}}{\partial x} dy = -\frac{d}{dx} \int_0^h \bar{u} dy + \bar{u}_h \frac{dh}{dx} \tag{15}$$

where $\bar{u}_h$ is the horizontal velocity at the surface in the x direction. By inserting Eq. 14 into 15, one obtains:

$$\bar{v}_h = \bar{u}_h \frac{dh}{dx} \tag{16}$$

Eq. 16 shows that $dh/dx > 0$, i.e., a decelerating flow generates $\bar{v}_h > 0$, but $dh/dx < 0$ or an accelerating flow yields the negative $\bar{v}_h$. Therefore, the vertical velocity in the main flows can also generate vertical velocity. Its interaction with groundwater can be obtained as $V_b = V_g + V_s$, or the groundwater velocity at the bed is jointly caused by Darcy velocity V_g and the vertical velocity caused by the main flow, V_s. Even the velocity on the solid–liquid interface may be very small, its importance for sediment transport should not be underestimated [3], and Eq. 16 shows that the vertical velocity has similar amplitude like the secondary current, i.e., about 1% of mean velocity.

Julien [5] replaced the Reynolds number in Shields' diagram by dimensionless particle diameter:

$$d_* = \left[\frac{\rho_s - \rho g d_{50}^3}{\rho} \frac{}{\nu^2} \right]^{1/3} \tag{17}$$

Similarly, d_* needs modification by introducing the apparent density with the following form:

$$d'_* = \left[\frac{\rho'_s - \rho}{\rho} \frac{g d_{50}^3}{\nu^2} \right]^{1/3} \tag{18}$$

Inserting Eq. 7 into Eq. 18, one has

$$d'_* = \left[\frac{\rho_s - \rho}{\rho} (1 - Y)^2 \frac{g d_{50}^3}{\nu^2} \right]^{1/3} \tag{19}$$

or

$$\frac{d'_*}{d_*} = (1 - Y)^{2/3} \tag{20}$$

Therefore, the empirical equation of Shields curve by Yalin and Silva [31] can be modified with the following form:

$$\tau'_* = 0.13 d'^{-0.392}_* \exp\left(-0.015 d'^2_*\right) + 0.045\left[1 - \exp\left(-0.068 d'_*\right)\right] \tag{21}$$

For the fall velocity, many empirical equations are available in the literature. Julien [5] related c_d in Eq. 4 with the particle diameter d_* and obtained the following empirical equation:

$$\frac{\omega d_{50}}{\nu} = 8\left[\sqrt{1 + 0.0139 d^3_*} - 1\right] \tag{22}$$

The incipient motion in uniform flows has been extensively investigated, but no one investigates the influence of vertical velocity on incipient motion, probably because this vertical flow may not large enough to induce discernible seepage, thus it is useful to estimate V_b using some measured parameters. The depth-average vertical velocity can be determined by,

$$V = U\frac{dh}{dx} \tag{23}$$

where U, the average streamwise velocity, and both U and dh/dx are measurable parameters, thus Eq. 23 is convenient to use.

The vertical velocity is jointly induced by either the groundwater or the surface variation, the joint effect can be assumed as the proportional V and the nominal seepage velocity V_s, i.e., $\lambda V + \lambda_s V_s$, or:

$$V_b = \frac{\lambda V + \lambda_s V_s}{1 - \varepsilon_0} \tag{24}$$

where λ and λ_s are the coefficients to relate V_b with the mean vertical velocity V and nominal seepage velocity (V_s) defined by Darcy ($V_s = ki$, k = hydraulic conductivity, i = hydraulic gradient), ε_0 = porosity of granular materials.

Generally in laboratory flumes, the second term of Eq. 24 is negligible (i.e., $V_s = 0$), but in natural streams both the river flow and underground water flow can generate the velocity at the river bed, thus two terms co-exist in Eq. 24.

4. Re-analysis of data on the original shields diagram

To verify whether Eq. 21 is applicable to non-uniform flows, 329 data points are comprehensively compiled [13, 17, 23, 29, 32–40]. The hydraulic conditions of the used data are summarized in **Table 1**, and the experimental conditions are briefly outlined as follows:

Neil conducted his experiments in a flume 0.9 m wide and 5 m long by using sands with different particle sizes and densities [32]. Among the data sets, 11 data points are obviously above the Shields curve. White collected his data from a recirculating flume 6 m long and 0.3 m wide, uniform sediment was used with diameter between (0.016–2.2) mm [33]. The experimental datasets by Everts included 35 runs with size d_{50} from 0.127 to 1.79 mm and specific gravity of 2.65, and 11 runs having d_{50} from 0.09 to 0.18 and specific gravity of 4.7 [34]. **Figure 4** shows that almost all his data points are located below the Shields' prediction. Carling's data [35] were collected from a narrow natural stream and in a broad stream. Graf and Suszka measured the critical shear stress in a flume 16.8 m long, 0.6 m wide and 0.8 m high, gravel sediment with uniform size was used [23]. Shvidchenko and Pender [36] used a flume to study the effect of relative depth on the incipient motion of coarse uniform sediments. Gaucher et al.'s [40] experiments were conducted in a horizontal, rectangular glass walled flume with dimensions of 6 m long, 0.5 m wide and 0.7 m deep, different types of non-cohesive materials were used ranged from d_{50} = 0.91 to 4.36 mm. Cheng and Chiew [30] investigated the influence of upward seepage on the critical conditions of incipient motion, the experiments were conducted in a horizontal flume 7.6 m long, 0.21 m wide and 0.4 m deep, with particle sizes of d_{50} = 0.63, 1.02 and 1.95 mm, and the seepage velocity (injection) was measured with a range between (0–0.0138) m/s. They found that the upward seepage reduces significantly the critical shear stress required by Shields curve. Kavcar and Wright [38] conducted experiments in a 7.5 m long, 0.6 m wide flume with both injection and suction seepage using sediment particle of d_{50} =0.16, 0.5 and 1.2 mm and the observed value of seepage velocity, i.e. V_s is range between (−0.0026–0.00223) m/s. Liu and Chiew [29] examined the critical shear stress for sediment with d_{50} = 0.9 mm subject to downward seepage with velocity between (−0.00314–0) m/s. Their glass-sided flume was 30 m long, 0.7 m wide and 0.6 m deep, they observed that the upward seepage (injection) decreases the critical shear velocity while the downward seepage (suction) increases it.

Nineteen flume experiments from Sarker and Hossain [37] are also included in **Figure 4**. They investigated the initiation of sediment motion under non-uniform sediment mixtures. Afzalimhr et al. [13] conducted experiments to investigate the effect of non-uniformity of flow on the critical shear stress in a channel (14 m long, 0.6 m width and 0.5 m depth), the sediment size of d_{50} = 8 mm was used for their observation. Different from Lamb et al's [24] prediction, their experimental data reveal that the value of critical shear stress is smaller than Shields' prediction by at least 50%. Similarly, Emadzadeh et al. [39] conducted experiments in accelerating and decelerating flow conditions, his flume was 14 m long, 0.6 m wide and 0.6 m deep. The sediment size used were d_{50} = 0.8 and 1.3, 1.8 mm for a total of 72 data sets. The decelerating/accelerating flows were obtained by adjusting negative and positive bed slope (±0.7%, ±0.9%, ±1.25% and ± 1.5%). It is found that the critical shear stress and Shields parameter for incipient motion in accelerating flow are higher than those predicted by Shields in uniform flow while their values in decelerating flow are considerably lower than that in accelerating flow.

These data mentioned are plotted in **Figure 4**, where the observed critical shear stress highly deviates from the standard Shields curve. All has been noticed and

Researchers	d_{50} (mm)	S	h (m)	U (m/s)	u_* (m)	τ_c'/τ_c	ω (m/s)	No. of data points	remark
Neil (1967)	5–29.1	0.01	0.03–0.192	0.28–0.35	0.029–0.165	0.685–1.18	0.13–0.62	59	uniform
Gaucher et al. (2010)	0.91–4.36	0.01	0.125–0.14	0.29–0.56	0.021–0.038	0.405–0.907	0.105–0.245	6	Uniform
Carling (1983)	62	0	0.213	0.163	0.141	—	0.94	2	uniform
	77	0	0.226	0.124	0.315		1.05		
Shvichenko & Pender (2000)	1.5–12	0.0019–0.0287	0.002–0.65	0.1–1.07	0.026–0.1157	—	0.14–0.41	21	uniform
White (1970)	0.016–2.2	0.02	0.02–0.07	0.0018–0.232	0.0062–0.045	—	0.00023–0.174	26	Non-uniform
Sarker & Hossain (2006)	0.64–1.02	0.00026–0.00063	0.089–0.214	0.3–0.59	0.019–0.024	—	0.084–0.1135	19	Non-uniform
Afzalimhr et al. (2007)	8	0.0075, 0.015	0.13–0.21	0.726–0.86	0.05–0.061	0.362–0.535	0.338	9	Non-uniform
Graf & Suszka (1987)	12.2,23.5	0.0075, 0.025	0.102–0.2	0.23–1.6	0.087–0.155	—	0.41–0.58	9	Non-uniform
Emadzadeh et al. (2010)	0.8, 1.3, 1.8 ±0.7, ±0.9, ±1.25, ±1.5		0.146–0.25	0.15–0.44	0.007–0.021	0.078–2.9	0.097–0.156	72	Non-uniform
Everts (1973)	0.09–1.79	0.005	0.0094–0.09	0.1312–0.38	0.018–0.043	0.39–1.79	0.007–0.156	35	Non-uniform
Liu & Chiew (2012)	0.9	0.01	0.12–0.14	0.28–0.35	0.0215	0.98–1.71	0.105	5	seepage
Cheng & Chiew (1999)	0.63–1.95	0.01	0.027–0.076	0.09–0.399	0.017–0.032	0.02–1.048	0.08–0.163	50	Seepage
Kavcar & Wright (2009)	0.16, 0.5, 1.2	0.01	0.23–0.29	0.23–0.412	0.013–0.022	0.67–1.84	0.019–0.124	16	seepage

Table 1.
Summary of experimental conditions by previous researchers.

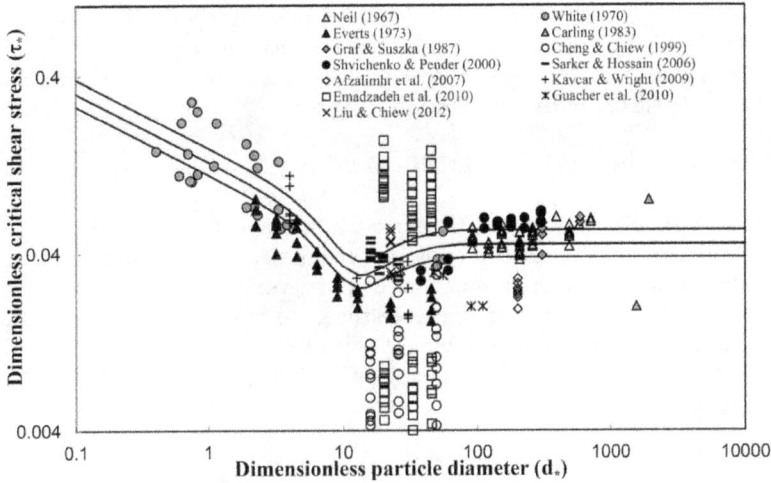

Figure 4.
Measured critical shear stress versus d∗ *and its comparison with shields curve or Eq. 21 at Y = 0 (i.e., uniform flow) and 100% error band.*

commented by many researchers [4, 24]. The consensus is that this discrepancy cannot be simply attributed to measurement errors or methodological bias. In **Figure 4**, the three lines are the Eq. 21 ($Y = 0$) ±100% error band.

5. Dependence of critical shields stress on channel slope

Many researchers have noticed that high channel's slope can cause the deviation of data from the Shields curve. For example Chiew and Parker [17] proposed that

$$\frac{\tau'_*}{\tau_*} = \cos\varphi\left(1 - \frac{\tan\varphi}{\cos\theta}\right) \tag{25}$$

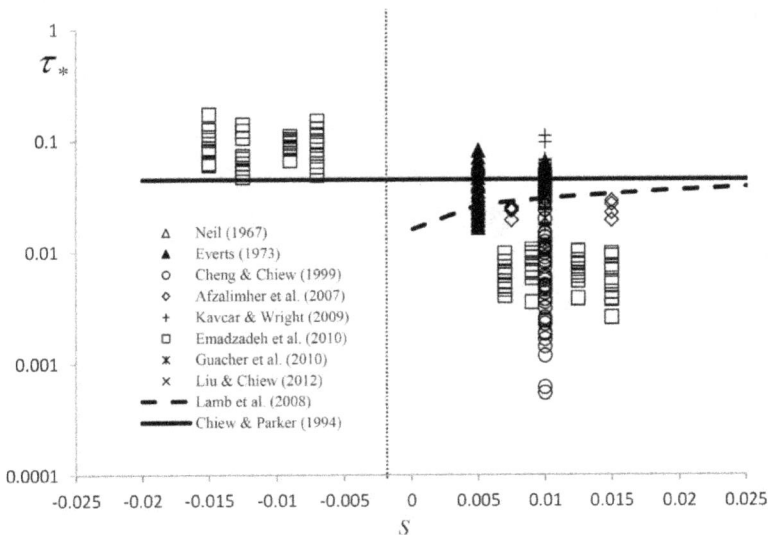

Figure 5.
Dependence of critical shear stress on the channel slope.

where ϕ = angle of streamwise bed slope, θ = angle of repose. Eq. 25 shows that the Shields number decreases with the increase of channel slope.

However, the formula given by Lamb et al. (2008) shows that the steep channel has a higher Shields number with the following form:

$$\tau'_* = \exp\left[0.0249X^4 + 0.107X^3 + 0.199X^2 + 0.476X - 3.57\right] \qquad (26)$$

where $X = 0.407ln(142\,S)$, and the slope S is in the regime $10^{-4} < S < 0.5$.

Figure 5 demonstrates the comparison of the measured data from **Table 1** and Eqs. 25 and 26. Obviously these equations do not agree the data points well. The measured τ_* could be largely different even the same type of sediment and channel slope are used. Therefore, the invalidity of Shields prediction cannot be simply explained by the dependence of channel slope, and there are some physics inside for the discrepancy.

6. Seepage on critical shields stress

Figure 5 demonstrates that for the same particle size in the same channel slope, the data points behave largely different, which cannot be explained by any existing theory. Beyond other factors, Eq. 24 shows that the scatter could be induced by either groundwater or the main flow's non-uniformity, or both of them. The effect of seepage on the critical shear stress is discussed first, the experimental data [29, 30, 38] are showed in **Figure 6**.

The modified Shields number in Eq. 8 (i.e., that with seepage) will be the same as that obtained from the Shields curve if one uses both the apparent sediment density and the apparent critical shear stress (that with seepage), i.e.

$$\tau'_* = \frac{\tau'_c}{(\rho'_s - \rho)gd_{50}} = \frac{\tau_c}{(\rho_s - \rho)gd_{50}} \qquad (27)$$

Using Eq. 7, one obtained the ratio of critical shear stresses with/without V_b in the following form:

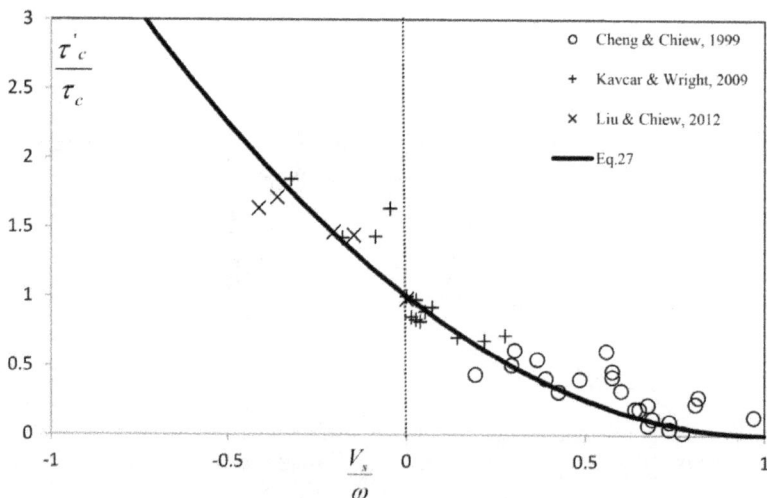

Figure 6.
Comparison of measured and predicted critical shear stress subject to seepage.

$$\frac{\tau_c'}{\tau_c} = (1 - Y)^2 \tag{28}$$

Figure 6 shows the critical shear stress predicted by Eq. 28 and the empirical factor λ_s is found to be 8.5, the experimenters determined the critical shear stress without seepage using the Shields curve. The good agreement between the measured and predicted critical shear stress indicates that the introduction of apparent sediment density is acceptable.

Similarly, local scour by large vortices (e.g., scour holes around bridge piers) is not caused by higher velocity or higher boundary shear stress, but the upward velocity Y. The mechanism is similar to the helicopter whose rotor blades generate the upward velocity by "vortices". Consequently, low water pressure induces the seepage or upward velocity, large particles like stones/helicopter can be lift. One can easily infer the relationship between the upward velocity and "vortices" in front of an electricity fan. Likewise, by observing how tornadoes damages large particles like cars, houses on surface, one can easily concluded that the upward velocity or lift force is the cause, by no means the shear force.

7. Effect of non-uniformity of flow on the critical shear stress

Figure 4 shows that the Shields' curve could be totally invalid sometimes, these noticeable deviations imply that the non-uniformity of flow could affect the predictability of Shields curve, for example, Afzalimhr et al.'s [13] data points locate below the curve when the flow was decelerating, Emadzadeh et al.'s data points [39] were far from the Shields' prediction, and his data points were obtained from both decelerating and accelerating flows. Hence, the large deviations from Shields curve shown in **Figure 4** can be used to verify Eq. 24, i.e., the vertical velocity induced by flow's non-uniformity is responsible for the invalidity of Shields curve.

To confirm whether the invalidity of Shields curve is caused by the non-uniformity of flow, the data without seepage in **Table 1** are used, and the water depth variation dh/dx is calculated using the following formula:

$$\frac{dh}{dx} = \frac{S - S_f}{1 - U^2/gh} \tag{29}$$

where dh/dx is the water depth's variation, S and S_f are the bed and energy slopes, respectively. Manning coefficient (n) can be assessed using the Strickler's formula:

$$n = \frac{d_{50}^{1/6}}{21.1} \tag{30}$$

The energy slope S_f in Eq. 29 can be determined from the Manning equation using the hydraulic radius R, i.e.,

$$S_f = \frac{n^2 U^2}{R^{4/3}} \tag{31}$$

In **Table 1**, the calculated dh/dx could be either negative or positive and the data points in **Figure 4** are included and replotted in **Figure 7**, where the data point is represented by the sign "+" if the obtained dh/dx is positive, otherwise the data point is marked by "-" for all negative dh/dx cases. **Figure 7** clearly shows that nearly all

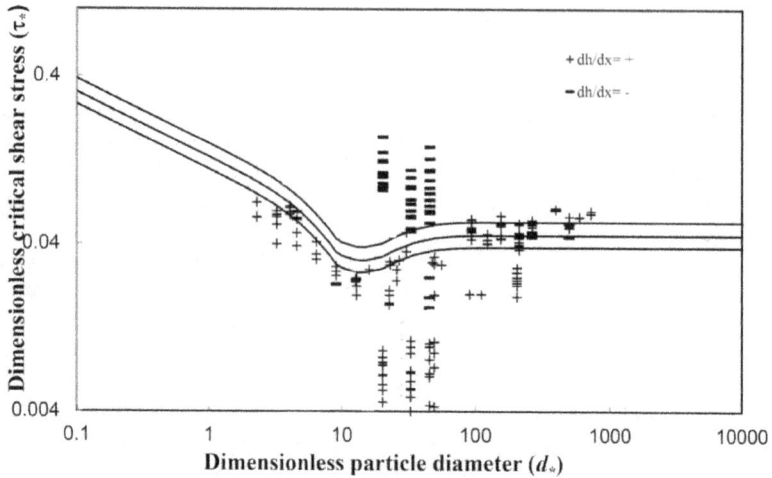

Figure 7.
*The variation of water depth dh/dx has different values based on the influence of vertical velocity on the initial motion, where (−0.024< **dh/dx** <0.0526) for all data sets from **Figure 3**.*

data points above the Shields curve have "-" signs, indicating the flows were accelerating, whilst almost all data points below the Shields curve have the sign of "+", or decelerating. Therefore, the non-uniformity of flow can play an important role for the deviation of measured critical shear stress from the Shields curve. **Figure 7** reveals that the presence of vertical velocity is one of the main causes responsible for the deviation of observed critical shear stress from the Shields curves for these data, the accelerating flow enhances particles' stability, and decelerating flow enables sediment's mobility. In **Figure 7**, the calculated positive dh/dx ranges from 0.000237 to 0.0526 $and − 0.024$ to −0.00073. It remains necessary to investigate whether the higher dh/dx has the higher deviation, and its analysis is shown below.

8. Modification of shields diagram

To examine whether data points without seepage shown in **Figure 4** can be expressed by Eq. 21, we can analyze the datasets without artificial seepage or with negligible groundwater effects, only those data are analyzed in which V_b is caused by the non-uniformity of flow in the main flow. Therefore, Eqs. 23 and 24 can be simplified as follows:

$$Y = \frac{V_b}{\omega} = \frac{\lambda U}{(1 − \varepsilon_0)\omega}\frac{dh}{dx} \tag{32}$$

Experiments [13, 34, 39, 40] are analyzed first. They reported that their measured critical shear stress is lower than Shields' prediction. Besides, the datasets [32, 39] are examined; they claimed that higher values of critical shear stress were observed.

In these studies, the experimental data sets from non-uniform flows are plotted in **Figure 8** where the empirical factor λ is found to be 8.5 for both decelerating and accelerating flows. The comparison of the predicted and measured critical shear stress in **Figure 8** shows that the agreement is reasonably good. Better agreement can be obtained if λ is calibrated as a function of sediment gradation and shapes, turbulence. Here, the assumption is that sediment particle size is uniform and can be represented by d_{50}.

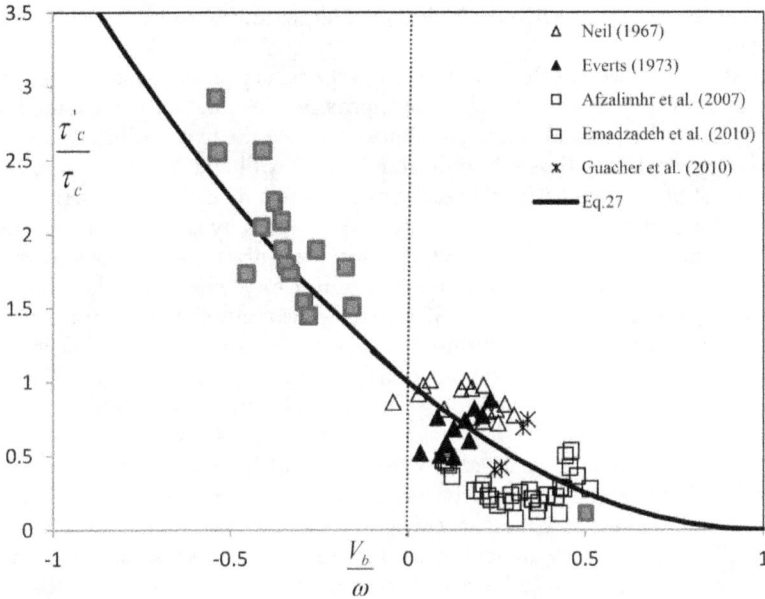

Figure 8.
Comparison of experimental results on threshold condition without seepage with Eq. 32.

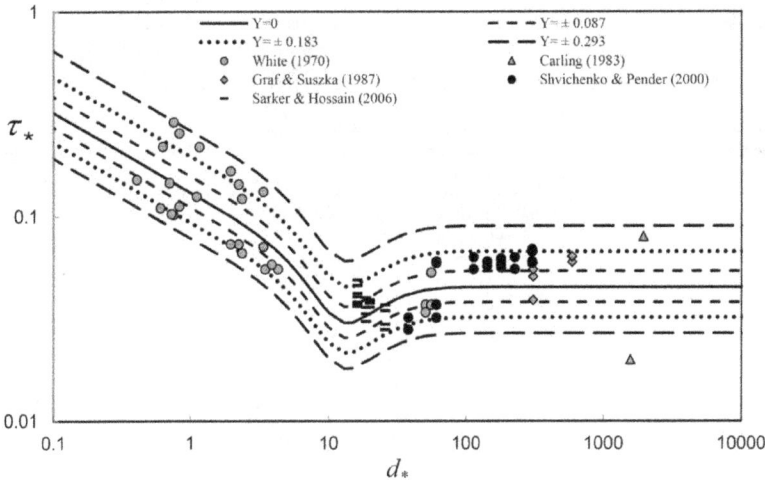

Figure 9.
Influence of vertical velocity on critical shear stress, the solid line is the original shields curve (or Y = 0) and other lines are calculated from Eq. 21 with different Y.

Figure 9 shows the comparison of measured and the predicted critical shear stress for the datasets [23, 33, 35–37]. Obviously, the observed critical shear stress largely deviates from the solid line, i.e., Shields curve ($Y = V_s/\omega = 0$), all data points can be covered by Eq. 9 or 10 when the parameter Y is used. In other words, **Figure 9** suggests that the scatter might be explained by variation of Y.

9. Discussion on slope's influence

As mentioned, some researchers have found the dependence of the critical shear stress on the channel slope, but it is still an open question about the validity of

Shields curve, especially when the bed slope is large, thus it is worthwhile to discuss this dependence.

This study reveals that the deviation from the Shields curve could be caused by the vertical velocity, the Shields curve is approximately valid only when the flow is uniform, when the vertical velocity is almost zero. As the true uniform is very rare in laboratory or nature, thus it is understandable why Shields curve is invalid to express most of observed critical shear stress. Hence, one needs to answer whether the dependence of τ_* on the channel slope is also caused by the flow's acceleration.

Eqs. 23 and 24 show that in almost all cases, there always exists the vertical velocity caused groundwater and flow's non-uniformity. Therefore, the widely observed dependence by Lamb et al. [24] may be also caused by the parameter Y ($\neq$ 0). Obviously, they assumed that the data used in their analysis were collected from uniform flows, thus these data can be used to compare the data with the Shields curve, and conclusion of the slope-dependence can be drawn. It is useful to examine this assumption by checking whether Lamb et al's data [24] are observed from uniform flows. Their data are listed in **Table 2**, in which only the laboratory data are included as their field data were certainly collected from non-uniform conditions. The last column of **Table 2** shows the length of flumes, and from it one can see that almost half of the flumes were less than 10 m. Kirkgöz and Ardiçlioğlu [41] measured the minimum length to form a uniform flow in a flume and found that a channel should be longer than 10 m as there is a transition zone from non-uniform flow to uniform flow. Even for those data from flumes longer than 10 m, the flow still could be non-uniform also when the parameter dh/dx is calculated using Eq. 6. Paola and Mohrig [42] suggest that uniform flow can only be assumed when the channel length is longer than h/S, if the water depth is 0.1 m, and slope is 1‰, this means that the channel length should be longer than 100 m. Therefore, one can conclude that it is likely that the data were generated in non-uniform conditions,

Researchers	d_{50} (mm)	τ_*	R_*	Flume length (m)
Neil (1967)	6.2, 8.5, 10.6, 20, 23.8, 29.1, 5, 16, 6.4	0.04–0.06	184.3–4800	5
Paintal (1971)	7.95, 2.5	0.05, 0.05	638, 112	15
Everts (1973)	3.57, 1.79, 0.895, 0.508, 0.395, 0.254, 0.127, 0.18, 0.09	0.018–0.07	1.3–162	16.8
Ashida & Bayazit (1973)	22.5, 12, 6.4	0.0386–0.1178		20
Fernandez Luque & Van Beek (1976)	0.9, 1.5, 1.8, 3.3	0.021–0.047	12–127	8
Ikeda (1982)	0.42 1.3	0.02 0.047	8.7 72	4
Graf & Suszka (1987)	12.2, 23.5	0.05–0.07	800–5000	16.8
Wilcock (1987)	1.83 1.83 0.67 5.28	0.03 0.036 0.023 0.037	61 12 332 115	23
Wilcock & Mcardell (1993)	5.3	0.02	219	7.9
Shvidchenko & Pender (2000)	1.5, 2.4, 3.4, 4.5, 5.65, 7.15, 9, 12	0.025–0.065	40–2000	6.5

Table 2.
Previously reported data selected from lamb et al. (2008).

this may lead to the different interpretation of the dependence of critical shear stress on the channel slope. **Figure 3** shows the accelerating flows in steep channels, thus it is likely that the downward velocity increases particles' stability.

Chiew and Parker's data [17] is used as an example, their observation is opposite to Lamb et al's prediction [24], they also found the dependence of critical shear stress on the channel slope based on their own data. In their experiments, the channel slope was specially adjusted from -10° to 31°, their channel lengths used were 4 m and 2 m only. Obviously, their experiments were conducted in the non-uniform flow conditions as the $2 \sim 3$ m length is too short to form a uniform flow. In other words, both conclusions drawn by Lamb et al. and Chiew and Parker [17, 24] are not very convincing as they did not check the parameter of dh/dx, and the data they used may be generated from non-uniform conditions.

10. Conclusions

This paper investigates why the observed critical shear stress widely deviates from the Shields curve, its discrepancy or validity could be caused by many factors like sediment shapes, gradation, measurement errors, turbulence and channel-bed slopes. However, this study reveals that the vertical motion also plays an important role, and the vertical velocity could be induced by non-uniformity of flow and seepage turbulence alike. After re-examining 329 data points from the literature, the following conclusions can be drawn:

1. The upward velocity increases sediment mobility and downward velocity increases sediment stability. The mobility or stability can be equivalently expressed by its apparent sediment density which is able to eliminate the effect of vertical velocity as shown in Eq. 7. This shifts a dynamic problem into a simplified static problem.

2. There exists vertical velocity on the channel bed and this vertical velocity could be induced by seepage or non-uniformity of flow, similar to the secondary currents, the small vertical velocity's influence on sediment incipient should not be underestimated. The joint effect is expressed by Eq. 24. For non-uniform flow, the sediment tends to move in decelerating flows, but it becomes more difficult to move in accelerating flows.

3. The Shields curve is valid only when the flow is nearly uniform, but a general Shields curve can be obtained by introducing the apparent sediment density, thus the modified Shields curve could be extended to express complex flows, this modified relationship for critical shear stress has been established.

4. A new parameter Y can be used to express the influence of non-uniformity of flow or seepage, this parameter should be included in the models of sediment transport. According to available experimental data in the incipient motion in non-uniform flows or in the seepage cases, good agreements between the measured and predicted values can be achieved if Y is included in the existing model, but more research is needed to determine the coefficients λ_s and λ in Eq. 24, they could be a function of sediment gradation and shapes, and turbulence.

All in all, high horizontal motion can make a plane (a big particle) to fly, high vertical velocity can also make the same particle called helicopter to fly.

Two mechanisms are totally different. It is wrong to ascribe all sediment transport phenomena to the horizontal motion only, without considering the vertical motion.

Notations

b = channel width
C_d = drag coefficient;
d_{50} = median size of sediment particles;
F_{vb} = force induced by the vertical velocity;
g = gravitational acceleration;
h = water depth;
i = hydraulic gradient;
k = hydraulic conductivity;
n = Manning coefficient;
Q = discharge;
R_* = Reynolds number;
S_f = energy slope.
U = mean velocity;
u_* = shear velocity;
u_{*c} = critical shear velocity $(\tau_c = \rho u_{*c}^2)$;
$\bar{u}$ and $\bar{v}$ = time-averaged velocity in the streamwise and vertical directions;
$\bar{u}_h, \bar{v}_h$ = horizontal and vertical velocities at the surface;
V = vertical velocity;
V_b = vertical velocity at the bed;
V_s = nominal seepage velocity at the bed;
$X = 0.407\ln(142S)$;
y = distance normal to the wall;
$Y = V_b/\omega$
ε_0 = porosity of granular materials
θ = angle of repose.
λ and λ_s = coefficients;
ν = kinematic viscosity;
ρ = fluid density;
ρ_s = sediment density;
ρ_s' = apparent density of sediment;
τ = boundary shear stress;
τ_c = critical boundary shear stress;
τ_* = Shields number;
τ_*' = modified Shields number subject to vertical velocity;
ϕ = angle of streamwise bed slope,
ω = particle fall velocity;
ω' = net falling velocity subject to vertical velocity.

Author details

Shu-Qing Yang* and Ishraq AL-Fadhly
School of Civil, Mining and Environmental Engineering, University of Wollongong, NSW, Australia

*Address all correspondence to: shuqing@uow.edu.au

IntechOpen

References

[1] Yang, C.T. (1996). Sediment transport: theory and practice. McGraw-Hill, Sydney.

[2] Shields, A. (1936). Application of similarity principles, and turbulence research to bed-load movement, California Institute of Technology, Pasadena (translate from German).

[3] Francalanci, S., Parker, G. and Solari, L (2008). Effect of seepage – induced nonhydrostatic pressure distribution on bed-load transport and bed morphodynamic. Journal of Hydraulic Engineering, vol. 134, no. 4, pp. 378–389.

[4] Buffington, J.M. and Montgomery, D.R. (1997). A systematic study of eight decades of incipient motion studies with special reference to gravel-bedded rivers. Water Resources Research, vol.33, no.8, pp. 1993–2029.

[5] Julien, P. Y. (1995). Erosion and Sedimentation. Cambridge University Press.

[6] Kramer, H. (1935), Sand mixtures and sand movement in fluvial models, ASCE Transactions, 100(1909), 798–838.

[7] Beheshti, A.A., B. Ataie-Ashtiani, (2008). Analysis of threshold and incipient conditions for sediment movement, Coastal Engineering, 55(5), Pages 423–430,

[8] Wiberg, P.L., and Smith, J.D. (1987). Calculations of the critical shear stress for motion of uniform and heterogeneous sediments, Water Resource Research, vol.23, pp.1471–1480.

[9] Wilcock, P.R. (1987). Bed-load transport in mixed-size sediment, PhD. Thesis, MIT, Cambridge.

[10] Johnston, C.E., Andrews, E.D., and Pitlick, J. (1998). In situ determination of particle friction angles of fluvial gravels, Water Resource Research, vol.34, no. 8, pp. 2017–2030.

[11] Kirchner, J.W., Dietrich, W.E., Iseya, F. and Ikeda, H. (1990). The variability of critical shear stress, friction angle, and grain protrusion in water worked sediments. Sedimentology, vol.37, pp.647–672.

[12] Garde, R. J. and Ranga Raju, (1985). Mechanics of sediment transportation and alluvial stream problems. 2nd ed. (New York: Wiley).

[13] Afzalimhr, H., Dey, S. and Rasoulianfar, P. (2007). Influence of decelerating flow on incipient motion of a gravel-bed stream. Sadhana, vol.32, no.5, pp. 545–559.

[14] Iwagaki, Y. (1956). Fundamental study on critical tractive force. Trans. Jan. Soc. Civil Engineering, no. 41, pp. 1–21.

[15] Andrews, E.D. and Kuhnle, R.A. (1993). Incipient motion of sand-gravel sediment mixture. Journal of Hydraulic Engineering, vol. 119, pp. 1400–1415.

[16] Dey, S. and Raju, U. (2002), Incipient motion of gravel and coal beds. Sadhana, vol.27, pp.559–568.

[17] Chiew, Y. M. and Parker, G. (1994). Incipient sediment motion on non-horizontal slopes. Journal of Hydraulic Research, vol. 32, pp. 649–660.

[18] Andrews, E. D. (1994). Marginal bed load transport in a gravel-bed stream, Sagehen Creek, California. Water Resource Research, vol.30, pp. 2241–2250.

[19] Church, M, Hassan, M. A., and Wolcott, J. F. (1998). Stabilizing self-organized structures in gravel-bed streams. Water Resource Research, vol.34, pp. 3169–3179.

[20] Patel, P. L., and Ranga Raju, K. G. (1999). Critical tractive stress of non-uniform sediments. Journal of Hydraulic Research, vol.37, pp.39–58.

[21] Dey, S., and Debnath, K. (2000). Influence of stream-wise bed slope on sediment threshold under stream flow. Journal of Irrigation Draining Engineering, vol.126, pp.255–263.

[22] Mueller E R, Pitlick J and Nelson JM (2005). Variation in the reference Shields stress for bed load transport in gravel-bed streams and rivers. Water Resource Research. vol.41, W04006, (doi: 10·1029/2004WR003692).

[23] Graf, W.H. and Suszka, L. (1987). Sediment Transport in Steep Channels. Journal of Hydroscience and Hydraulic Engineering, vol.5, no. 1, pp.11–26.

[24] Lamb, M.P., Dietrich, W.E. and Venditti, J.G. (2008). Is the critical Shields stress for incipient sediment motion dependent on channel-bed slope? Journal of Geophysical Research, vol.113, F02008

[25] Yang S.Q. and J. W. Lee (2007). "Reynolds shear stress distributions in a gradually varied flow". Journal of Hydraulic Research, IAHR. 45(4), 462–471.

[26] Yang S.Q. and A. T. Chow (2008). "Turbulence structures in non-uniform flows", Advances in Water Resources, 31, 1344–1351.

[27] Yang S.Q. (2009). "Velocity distribution and wake-law in gradually decelerating flows". J. Hydr. Res., IAHR, 47(2), 177–184.

[28] Rao, A. R., and Nagaraj, S. (1999). Stability and mobility of sand-bed channel affected by seepage. Journal of Irrigation Draining Engineering, vol.125, no.6, pp.370–379.

[29] Liu, X.X. and Chiew, Y.M. (2012). Effect of seepage on initiation of cohesionless sediment transport. Acta Geophysica, vol. 60, no. 6, pp. 1778–1796, DOI: 10.2478/s11600-012-0043-7.

[30] Cheng, N-S and Chiew, Y-M (1999). Incipient sediment motion with upward seepage. Journal of Hydraulic Research, vol.37, no.5, pp665–681.

[31] Yalin, M.S. and Silva, A.M.F. (2001). Fluvial processes. IAHR, Delft, the Netherlands.

[32] Neill, C. (1967). Mean-Velocity Criterion for Scour of Coarse Uniform Bed-Material. International Association for Hydraulic Research, pp. 46–54.

[33] White, S J. (1970). Plane bed thresholds of fine grained sediment. Nature, vol.228, pp. 152–153.

[34] Everts, C.H. (1973). Particle overpassing on flat granular boundaries. Journal of the Waterways Harbors and Coastal Engineering Division, vol. 99, pp. 425–439.

[35] Carling, PA. (1983).Threshold of coarse sediment transports in broad and narrow natural streams. Erath Surface Processes and Landforms, vol.8, pp.1–18.

[36] Shvidchenko, AB. and Pender, G. (2000). Flume study of the effect of relative depth on the incipient motion of coarse uniform sediments. Water Resources Research, vol. 36, no.2, pp. 619–628.

[37] Sarker, LK and Hossain, MM (2006). Shear stress for initiation of motion of median sized sediment of no uniform sediment mixtures. Journal of Civil Engineering (IEB), vol.34, no. 2, pp. 103–114.

[38] Kavcar, P.C., and Wright, S.J. (2009). Experimental results on the stability of non-cohesive sediment beds subject to vertical pore water flux. Proc. World Environmental and Water

Resources Congress 2009: Great Rivers, vol. 342, pp. 3562–3571, DOI: 10.1061/41036.

[39] Emadzadeh, A, Chiew, YM and Afzalimehr, H. (2010). Effect of accelerating and decelerating flows on incipient motion in sand bed streams. Advances in Water Resources, vol.33, pp. 1094–1104.

[40] Gaucher, J, Marche, C and Mahdi, T-F. (2010). Experimental investigation of the hydraulic erosion of non - cohesive compacted soils. Journal of Hydraulic Engineering, vol.136, no. 11, pp. 901–913.

[41] Kirkgöz, M. S. and Ardiçlioğlu M. (1997). Velocity Profiles of Developing and Developed Open Channel Flow. Journal of Hydraulic Engineering, vol.123, no. 12, pp. 1099–1105.

[42] Paola, C., and D. Mohrig (1996), Paleohydraulics revisited: Paleoslope estimation in coarse-grained braided rivers, Basin Research, 8, 243–254.

Chapter 2

Formulae of Sediment Transport in Unsteady Flows (Part 2)

Shu-Qing Yang

Abstract

Sediment transport (ST) in unsteady flows is a complex phenomenon that the existing formulae are often invalid to predict. Almost all existing ST formulae assume that sediment transport can be fully determined by parameters in streamwise direction without parameters in vertical direction. Different from this assumption, this paper highlights the importance of vertical motion and the vertical velocity is suggested to represent the vertical motion. A connection between unsteadiness and vertical velocity is established. New formulae in unsteady flows have been developed from inception of sediment motion, sediment discharge to suspension's Rouse number. It is found that upward vertical velocity plays an important role for sediment transport, its temporal and spatial alternations are responsible for the phase lag phenomenon and bedform formation. Reasonable agreement between the measured and the proposed conceptual model was achieved.

Keywords: unsteady flow, shields number, rouse number, sediment transport, vertical velocity

1. Introdution

Sediment transport is the movement of solid particles driven by fluid like water or wind in rivers, lakes, reservoirs, coastal waters. Generally, in the real world the flow is unsteady like flood waves, tidal waves and wind waves, because steady and uniform flows are very rare in reality. Even so, it is understandable that sediment transport is first observed under well controlled conditions in laboratory, and then the data are collected to calibrate the models. These formulae are further examined using field data by assuming the laboratory flow conditions (generally steady and uniform flows) can be extended to rivers and coastal waters (generally unsteady and non-uniform).

In the literature, many formulae use the boundary shear stress τ ($=\rho ghS$) to express sediment discharge, like *Einstein* [1], *Meyer-Peter* and *Muller* [2], *Yalin* [3], *Engelund* and *Hansen* [4] and *Ackers and White* [5]. For example, the Meyer-Peter and Muller equation for the bed load and Engelund-Hansen formula for the total load have the following forms:

$$\frac{g_b}{\sqrt{(\rho_s/\rho - 1)gd_{50}}} = 8.0\left(\frac{\tau}{(\rho_s - \rho)gd_{50}} - 0.047\right)^{1.5} \qquad (1)$$

$$c_f\frac{g_t}{\sqrt{(\rho_s/\rho - 1)gd_{50}}} = 0.1\left[\frac{\tau}{(\rho_s - \rho)gd_{50}}\right]^{5/2} \qquad (2)$$

where g_b and g_t = bed-load and total load of sediment discharge per unit width, g = gravitational acceleration, d_{50} = median sediment size, ρ_s = sediment density, and ρ = water density, h = water depth, S = energy slope, c_f = friction factor which is constant in fully rough regime. The subscribes b and t denote the bed load and the total load. Eqs. (1) and (2) demonstrate that if d_{50}, ρ_s are constant, sediment discharge only depends on τ.

Alternatively, the mean velocity U was selected to represent the hydraulic parameter for sediment discharge or concentration like the *Velikanov's* [6] parameter, $U^3/(gh\omega)$. The WIHEE's [7] equation which has been widely used in China has the following form:

$$C = k_1 \left(\frac{U^3}{gh\omega} \right)^m \tag{3}$$

where C = sediment concentration, k_1 and m are empirical coefficient, ω = sediment settling velocity.

Besides the parameters U and τ alone, attempts have been made to correlate the sediment transport with the product of U and τ. Probably *Bagnold* [8] was the first one to do so, and it is known as the stream power (= τU). Likewise, the product of U and S, or the unit stream power US/ω was used by *Yang* [9]. *van Rijn* [10] selected u_*', the shear velocity related grains, in his equations, i.e., T, and d_*, they are

$$T = \frac{u_*'^2 - u_{*c}^2}{u_{*c}^2} \tag{4}$$

$$d_* = d \left[\frac{(\rho_s/\rho - 1)g}{\nu^2} \right]^{1/3} \tag{5}$$

where

$$u_*' = \frac{U}{2.5 \ln \frac{11h}{2d_{50}}} \tag{6}$$

where the critical shear stress $\tau_c = \rho u_{*c}^2$, ν = kinematic viscosity.

Yang and Tan [11] found that the shear velocity u_*' is responsible for transporting the sediment particles, *Yang* [12] defined the energy dissipation on sediment transport as $E = \tau u_*'$, and obtained the formula of sediment transport:

$$\overrightarrow{g_t} = \left(\frac{\rho_s}{\rho_s - \rho} \right) k \overrightarrow{u_*'} \left(\frac{E - E_c}{\omega} \right) \tag{7}$$

where the arrows represent the direction, i.e., sediment is transported in the same direction as the near bed flow if the flow directions of upper and lower layers are different, E_c (= ρu_{*c}^3), k is a constant (= 12.2) and insensitive to other hydraulic parameters like Froude number, Reynolds number, relative roughness and Rouse number [13].

Obviously, the hypothesis in all equations listed above is that the higher the streamwise parameters are (e.g., U, u_*', τ, E or US etc.), the more particles are transported [14]. However, this prediction is invalid in unsteady conditions [15, 16]. Tabarestani and Zarrati [17] reviewed the performance of existing formulae and concluded that in general, the sediment discharge under unsteady flow conditions cannot be predicted by these equations, because the streamwise parameters in the rising limb is much larger than that in the falling limb, but the measured sediment load yield during hydrograph rising limb is smaller than that in the falling limb. The highest g_t or C comes after the peak flowrate or velocity U, and the lag phenomenon has been widely observed and reported. The shear stress based theory has also been questioned by *Nelson et al.* [16] who observed from their experiment that the sediment flux increases even though the bed shear stress decreases.

Sleath [18] argued that when the "pressure gradient" is not small compared with the shear stress exerted by the flow, these equations need to be modified and a new S_1 *number* should be considerd for wave conditions, its definition is:

$$S_1 = \frac{\rho U \sigma}{(\rho_s - \rho)g} \tag{8}$$

where σ is the angular frequency of waves.

Alternatively Francalanci et al. [19] suggest using the pressure P to express the unsteadiness, but Liu and Chiew [14] and Cheng and Chiew [20] use the hydraulic gradient i in the sediment layer. The challenge also comes from the bursting phenomenon even in steady and uniform flows. It is found that the similar lag phenomenon exists in a bursting cycle [21, 22]. *Cellino and Lemmin's* [23] experiments demonstrate that the upward flow (or ejection) appears responsible for the threshold of particle movement, the entrainment and transport of bedload and lifting of sediment into suspension. This cannot be explained by the parameters of pressure P or hydraulic gradient i or seepage velocity.

It seems that there is a knowledge gap between the unsteady flows and sediment transport, a new parameter is needed to be developed to express the unsteadiness, thus the above phenomena can be explained. In this study, the induced vertical velocity V is selected to express the effect of unsteadiness on the sediment, an attempt is made to justify its suitability for sediment transport as well as the phenomena of phase-lag and bedform formation. The research objectives include:

1. to compare V with other parameters to express the force induced by unsteady flows;

2. to establish a simple connection between V_b in the sediment layer and V in the main flow;

3. to develop formulae to express critical shear stress, sediment discharge and Rouse number in unsteady flows;

4. to explain the mechanism of phase lags and bedform formation.

The chapter discusses the existence of vertical velocity in unsteady flows first, then the influence of vertical velocity on critical shear stress of sediment is analyzed, followed by its influence on sediment discharge and suspension concentration. Finally a comprehensive discussion is provided.

2. Theoretical consideration

Sediment transport is a joint result of streamwise and vertical motions of fluid. This joint effect can be seen from the definition of Shields number that is the ratio of forces in streamwise and vertical directions as noted by *Francalanci et al.* [19]:

$$\tau_* = \frac{\tau}{(\rho_s - \rho)gd} \tag{9}$$

where τ_* = Shields number. The numerator denotes the streamwise friction force and the denominator represents the vertical force, i.e., the net buoyant force of particle. Sediment starts to move at $\tau_* \geq \tau_{*c}$, the critical Shields number.

A simple wave model is shown in **Figure 1a** where a surface wave induces a vertical motion for the particles on the permeable bed. The surface wave is propagating in the research domain where the current velocity is U, the streamwise parameters like the point velocity, shear stress, pressure P and hydraulic gradient i in the soil are also modified. In this study, the induced vertical motion has been expressed by velocity at the interface is V_b. The relationship between the wave and its induced vertical velocity is shown in **Figure 1b**.

In **Figure 1**, the continuity equation of unsteady flows must be satisfied, i.e.,

$$\frac{\partial u}{\partial x} + \frac{\partial v}{\partial y} = 0 \tag{10}$$

where u and v are the streamwise and vertical time-averaged velocities in x and y directions. The vertical velocity can be determined from Eq. (10) as follows:

$$v = -\int \frac{\partial u}{\partial x} dy \tag{11}$$

In Eq. (11) the term $\partial u/\partial x$ is the gradient of streamwise velocity in x-direction, it is positive if the velocity becomes higher to downstream (accelerating), and

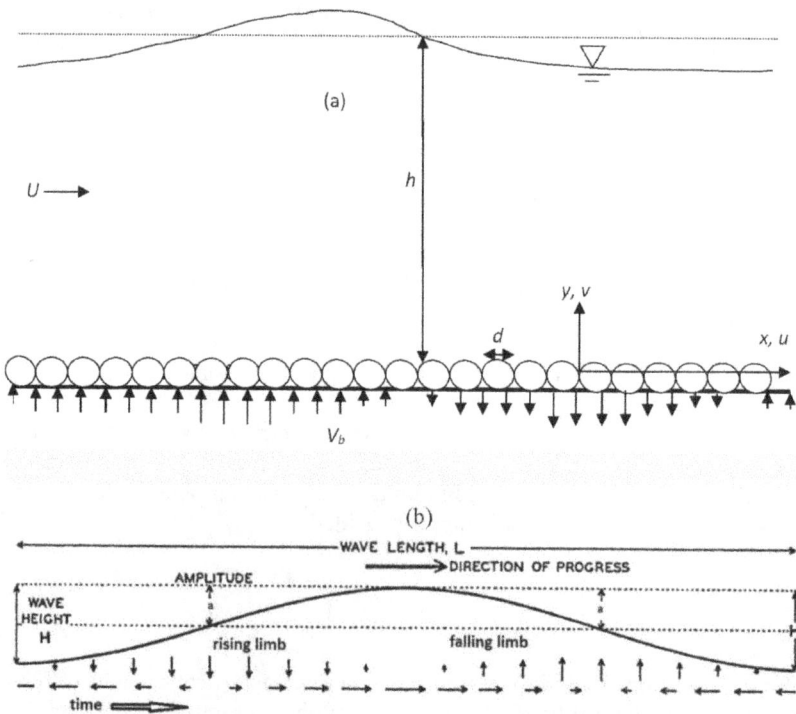

Figure 1.
(a) Schematic diagrams showing interaction of surface waves and induced and vertical motions at the sediment layer along x direction. (b) Definition of progressive wave and its induced vertical velocity at different time (x = constant).

negative if the fluid particles experience decelerating. Hence, the accelerating flow yields a negative or downward v, the decelerating flow generates an upward or positive v.

At the permeable boundary, the fluid velocity must meet the continuous boundary condition, i.e., $v_{(y=0^+)} = v_{(y=0^-)}$, or the velocity inside the sediment layer must be same as the velocity in the main flow at the interface. Thus it can be concluded that a downward velocity exists in the sediment layer when the main flow layer is accelerating, and an upward velocity appears when a flow is decelerating.

Generally speaking, the rising limb is the accelerating stage which induces a downward velocity, but the decelerating stage in ebb limb generates an upward velocity. In the real world, it is also possible that flows in both rising/falling limbs are accelerated as observed by *Song and Graf* [24], who used acoustic Doppler velocity profiles measured the vertical velocity in unsteady open channel flows, and found during the rising/falling limbs, "the measured vertical velocity are almost always negative, and this implies that the flows of the present experiments are accelerating ones". On the other hand, *Leng and Chanson* [25] used an acoustic Doppler velocimeter (ADV) measured the vertical velocity in tidal bores and found that the vertical velocity is always upward or decelerating in both rising and falling limbs. To simplify the discussion, this study only discusses the cases shown in **Figure 1b** and the waves' influence on parameters like q, U is assumed to be negligible.

The direction of vertical velocity can noticeably change the profile of Reynolds shear stress, streamwise velocity etc. [12, 26]. One of the examples is shown in **Figure 2**, *Kemp and Simons* [27, 28] measured the velocity profiles in a flume where the incident wave was set to propagate against or along the direction of the currents. The flow depth at the test section was kept at 200 mm for all tests. Regular waves were generated with a constant wave period of 1 second. The wave heights were 27.9 to 20.7 mm, the wave lengths were 1053 mm to 1426, respectively. Their results clearly show that the measured velocity is greater than log-law's prediction when waves opposite the current as the original uniform flow is decelerated by the waves from downstream, but less than the log-law's prediction when the waves to the currents as the original uniform flow is accelerated by waves from upstream. Existing research [26, 29] shows that in a turbulent flow the log-law is satisfied if the upward velocity in the main flow $V = 0$, but the measured velocity is higher than the log-law's prediction if $V > 0$ or upward velocity exists, and the maximum velocity is submerged if $V < 0$ (or downward velocity exists). Further investigation shows that a decelerating flow generates an upward velocity, but an accelerating

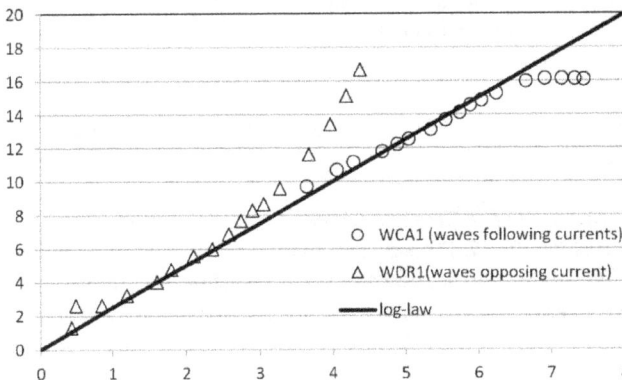

Figure 2.
Deviation of measured velocity from log-law by Kemp and Simons [21, 22].

flow induces downward velocity [30]. Therefore one can infer that in **Figure 2**, there exists an upward velocity for waves against a current or the waves make the current decelerated; but a downward velocity exists in the case of waves following current, which accelerates the water.

For sediment particles in **Figure 1**, the settling velocity ω in still water is determined by:

$$C_d \pi \frac{d^2}{4} \frac{\rho \omega^2}{2} = \pi \frac{d^3}{6} g(\rho_s - \rho)$$

(12)

where drag coefficient C_d depends on the Reynolds number Re ($= \omega d/\nu$) and $C_d = 0.45$ for large Reynolds number, i.e., Re >1000.

If a surface wave induces an upward velocity V_s in the preamble sediment layer, the net settling velocity is reduced to $\omega - V_b$. The reduction of settling velocity could be treated by altering its density from ρ_s to ρ_s' by assuming the particle's size remains unchanged, and the force balance equation is similar to Eq. (12) with the following form:

$$C_d' \pi \frac{d^2}{4} \frac{\rho(\omega - V_b)^2}{2} = \pi \frac{d^3}{6} g(\rho_s' - \rho)$$

(13)

From Eqs. (12) and (13), one can derive the following relationship:

$$\frac{\rho_s' - \rho}{\rho_s - \rho} = \alpha \left(1 - \frac{V_b}{\omega}\right)^2$$

(14)

where $\alpha = C_d'/C_d$ and $\alpha = 1$ are assumed to simplify the mathematical treatment. Eq. (14) shows that if V_b is upward, then $\rho_s' < \rho_s$, or the sediment particles become lighter in the "boiling" environment. If the upward $V_b = \omega$, Eq. (14) shows that the sediment density is similar to the water density $\rho_s' = \rho$. If the sediment particles are exercising the downward velocity (negative V_b), then the density $\rho_s' > \rho_s$, or the sediment behaves like heavy metals. As the decelerating velocity can generate upward velocity, it can be inferred that if the streamwise parameters keep almost unchanged, the sediment can be more easily transported in decelerating flows relative to the accelerating flows. In other words, the sediment particles become lighter in decelerating flows (or decelerating phase), but heavier in accelerating flows/phase. As Eq. (10) is also valid for turbulent velocity and wave conditions, then the conclusion can be extended to the bursting phenomenon or wave conditions where the accelerating/decelerating phases alternate randomly or regularly, thus these equations provide a general tool to analyze sediment transport.

3. Influence of unsteadiness on critical shear stress for incipient sediment transport

It is interesting to discuss how the waves affect the initiation of sediment movement. For an unsteady flow, the existing Shields diagram may be invalid to express the threshold sediment motion, due to the existence of vertical velocity caused by its unsteadiness. When the apparent sediment density is included in the Shields number, it has the following form:

$$\tau_*' = \frac{\tau_c'}{(\rho_s' - \rho)gd}$$

(15)

where τ_c' is the critical shear stress with vertical velocity. Inserting Eq. (14) into Eq. (15), one has:

$$\tau_*' = \frac{\tau_c'}{(\rho_s - \rho)gd} \left(\frac{\omega}{\omega - V_b} \right)^2 \tag{16}$$

Using Eq. (9), Eq. (16) can be rewritten as follows

$$\frac{\tau_*'}{\tau_*} = \left(\frac{\omega}{\omega - V_b} \right)^2 \tag{17}$$

Eqs. (16) and (17) generally express the relationship between the Shields number τ_*' with waves and the original Shields number τ_* without waves. It predicts that the original Shields number may significantly deviate from the Shields curve subject to wave conditions.

Eq. (15) includes the influence of the vertical velocity, it demonstrates that the upward velocity reduces particles' apparent density, thus the required critical shear stress will be also reduced. Whilst the downward velocity increases the apparent density, thus the required critical shear stress is higher. If the cases with/without vertical velocity are compared, the critical shear stress without waves τ_c and the critical shear stress with waves τ_c' have the following relationship:

$$\frac{\tau_c'}{\tau_c} = (1 - Y)^2 \tag{18}$$

and

$$Y = \frac{V_b}{\omega} \tag{19}$$

Eq. (18) shows that the critical shear stress τ_c' in unsteady flows. It should be stressed that for sediment incipient motion, V_b in Eq. (19) depends on the instantaneous maximum upward velocity for which the ejection of burst phenomenon, unsteadiness and others may jointly contribute. For flows shown in **Figure 2**, one can infer that the measured τ_c' is less than Shields diagram's prediction when the waves propagate against the current, but the τ_c' becomes larger than τ_c when the waves propagate with the current. The reason is that, the former generates an upward velocity in the decelerating flows, but the latter has a downward as it is an accelerating flow.

If the influence of small wave on the shear stress is negligible, the Y with small waves must be higher than the Y without waves. In such case, one can easily conclude from Eq. (18) that the τ_c' (with waves) must be always less than τ_c (without waves). In the literature, it seems that many researchers agree that the existing Shields diagram can be extended to the wave-current motion (i.e., [31, 32]). Till recently, few researchers like *Green and MacDonald* [33] found waves, not currents initiate sediment transport, their data show that "observed τ_*' never exceeded the theoretical dimensionless τ_*". It is well known that for large particles, the critical Shields number $\tau_* = 0.06$. They observed suspension at the same value of τ_* when waves are present in tidal flows, similar observations were reported by *Green and Coco* [34]. All of these observations can be easily explained by Eq. (18) when $Y \approx 1$.

It should be stressed that accelerating flows constrain sediment mobility from vertical point of view, but the higher velocity and shear stress in the rising limb

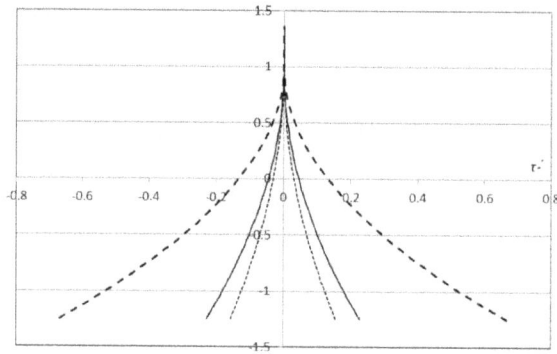

Figure 3.
Sediment incipient conditions in wave conditions, the required shields numbers depend on the veritocal motion, i.e., ±Y. based on Eq. (15), the calculated solid line (——) represents non-cohensive sediment in shields diagram $\tau_*' = 0.045$; the dotted line (·········) for $\tau_*' = 0.03$; and the dashed line (– –) for very fine sediment with $\tau_*' = 0.13$. Below these curves, particles remain static, above the curves particles are in mobile state.

promote sediment transport in the streamwise direction, therefore the complete effect of accelerating flows in the rising limb should include both shear stress and maximum Y. Likewise, the decelerating flow makes particles "lighter" in vertical direction, but the reduced shear stress makes particles to move "harder". Therefore, one need to justify the critical shear stress by considering both streamwise and vertical parameters.

Eq. (18) clearly demonstrates that the critical shear stress is jointly determined by the streamwise and vertical motions. The coexistence of streamwise/vertical motions results in the invalidity of Shields diagram which can be improved by Eq. (18) and shown in **Figure 3**, where the Shields number in the original Shields diagram is $\tau_* = 0.045$, 0.03 and 0.13 are calculated using Eq. (18). The region below the curves represents that the sediment is static, and above these curves is mobile. The calculated results show that when $Y \geq 0.7$, the sediment is mobile, for which the required shear stress is always zero.

4. Effects of vertical velocity induced by waves on sediment transport

As mentioned before that sediment transport is a joint effect of streamwise and vertical motions, the latter can be represented by the apparent sediment density. Therefore, Eq. (7) can be modified with the following way:

$$\vec{g}_t = \left(\frac{\rho_s'}{\rho_s' - \rho}\right) k \vec{u}_*' \left(\frac{E - E_c}{\omega - V_b}\right) \tag{20}$$

For sediment transport in waves conditions, the bed shear stress $\tau = \tau_w + \tau_{cu}$ and near bed velocity $u_b = u_w + u_{cu}$, where the subscripts w and cu refers to waves and currents. Yang [12] obtained the formula which agrees reasonably well with von Rijn's data in 1993, 1995 and 1999 for sediment transport when waves follow or oppose the currents, there are some angles between the direction of wave propagation and current, and waves are broken over a near shore bar, respectively. Even the best agreement has been achieved among the existing formulae, noticeable discrepancies imply that some mechanism of sediment transport by waves needs further investigations.

Eq. (20) shows that the direction of sediment motion is always the same as the near bed velocity. This is meaningful to specify the sediment moving direction in coastal waters where the direction of flow in up layer is often different from that in the bottom layer. Eq. (20) has the following simplified form [35]:

$$g_t = k \frac{\rho_s'}{\rho_s' - \rho} \tau_o \frac{u_*'^2 - u_{*c}'^2}{\omega'} \tag{21}$$

Inserting Eq. (12) into Eq. (21), one has:

$$g_t(Y) = k \left[\frac{\rho}{\rho_s - \rho} \left(\frac{1}{1 - Y} \right)^2 + 1 \right] \tau_o \frac{u_*'^2 - u_{*c}'^2}{\omega(1 - Y)} \tag{22}$$

Eq. (22) shows that sediment transport rate is jointly determined by the streamwise flow conditions (i.e., τ_o and u_*') and Y.

For the maximum over-the-wave-cycle horizontal wave-orbital speed at the bed U_b can be expressed by the wave height H and the wave period T, as both these govern the wave-orbital speed at the bed at any given water depth h. For linear waves, this is expressed as

$$U_b = \frac{\pi H}{T \sinh(k_o h)} \tag{23}$$

where the dispersion relationship gives:

$$\sigma^2 = g k_o \tanh(k_o h) \tag{24}$$

$\sigma = 2\pi/T$, k_o is the wave number and $k_o = 2\pi/L$, L = wave length.

It can be assumed that at the interfacial boundary, $v(y = 0^+)$ has the same magnitude order as U_b, and the vertical velocity at the sediment layer can be expressed as

$$V_b = \beta U_b \tag{25}$$

To evaluate the influence of vertical velocity on sediment transport rate, one can compare the sediment transport rate in two cases: with or without the vertical velocity induced by waves if τ_o remain unchanged. At $V_b = 0$, Eq. (22) becomes:

$$g_t(0) = k \frac{\rho_s}{\rho_s - \rho} \tau_o \frac{u_*'^2 - u_{*c}'^2}{\omega} \tag{26}$$

From Eqs. (22) and (26), one has:

$$\frac{g_t(Y)}{g_t(0)} = \frac{\rho}{\rho_s(1 - Y)^3} + \frac{\rho_s - \rho}{\rho_s(1 - Y)} \tag{27}$$

where $g_t = Cq$ and C = sediment concentration. For a current with very small waves, the influence of small waves on the discharge q is negligible, thus $g_t(Y)/g_t(0) \approx C(Y)/C(0)$.

Green [36] measured sediment concentration in an estuarine intertidal flat in New Zealand under very small waves. The wave height is less than 10 cm, and wave period ranges from 1.0–1.8 s. The measured data shows that sediment concentration in the rising tide is not very high, the highest concentration is always appear in the ebb tide. Eq. (11) may provide an explanation when the rising tide is assumed to be accelerating and the falling tide is decelerating. A downward velocity is generated

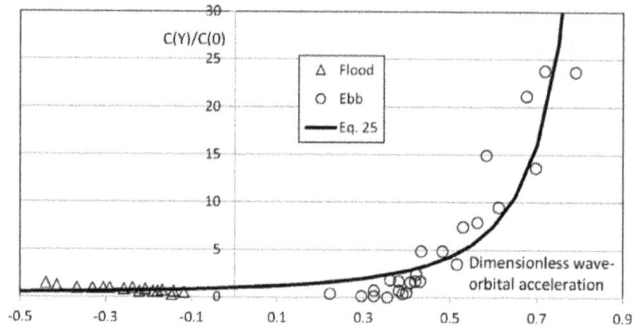

Figure 4.
Measured sediment concentration normalized by C(0) = 5 mg/L versus the wave-orbital acceleration normalized by 23 cm/s². The raw data were deprived from Green [17], the acceleration in flood limb was set to negative and the acceleration in ebb stage was set to be positive. After this transformation, the obtained data can structurally match Eq. (27), implying the connection between the dimensionless parameters Y and the wave-orbital acceleration.

the rising tide, which has the same effect on sediment as the particle's density becomes heavier. But during the falling limb or low tide, the particles become lighter, so the concentration becomes higher in this stage as shown in **Figure 4**.

In their analysis, *Green* [36] found that the "wave-plus-current-stress" theory provides poor agreement with their data. But the "wave-orbital speed" theory performed the best at predicting the incipient motion and suspension. They found a strong relationship between the measured sediment concentration and the wave-orbital acceleration a_0 which is defined as:

$$a_0 = \frac{1}{n} \sum_{i=1}^{n} (|U_{ZDC+}| - |U_{ZDC-}|)_i / T_{ZDCi} \qquad (28)$$

where n = the number of zero-down crossing waves in the burst, U_{ZDC+} is the maximum zero-down crossing current excursion in the positive direction from its average velocity. U_{ZDC-} is the maximum zero-down crossing current excursion from the mean velocity in the negative direction, T_{ZDC} is the period for the events.

Figure 4 shows a plot $C(Y)/C(0)$ versus Y ($=a_0/23$). Green [36] plotted his measured concentration in mg/L against a_0 using Eq. (28), in which the wave period is almost constant, thus the acceleration a_0 in Eq. (28) is actually the velocity. In **Figure 4**, the averaged concentration in the flood stage is used as C(0) and C (0) = 5 mg/L. It is found that data points match Eq. (27) very well when the acceleration a_0 is normalized by 23 cm/s² that is not clear the reason. In the calculation, the sediment and seawater densities are 2650 and 1025 kg/m³, respectively.

It can be seen that the sediment transport rate can be significantly promoted by an ebb tide, if the upward velocity is 75% of settling velocity (Y = 0.75), then the predicted sediment transport rate can be increased to 27 times of $g_t(0)$. **Figure 4** also shows that the sediment transport rate is slightly reduced if a downward flow exists. If Y = −0.5, then the sediment transport rate will be reduced to 1/2 of $g_t(0)$, this transport rate is achieved as the particles becomes "heavier".

5. Sediment suspension by tidal waves

The governing equation of suspended concentration can be derived from the continuity equation of solid-phase in the following form [37].

$$\frac{\partial c}{\partial t} = \frac{\partial\left(cu + \overline{c'u'}\right)}{\partial x} + \frac{\partial\left(cv + \overline{c'v'} - c\omega\right)}{\partial y} + \frac{\partial\left(cw + \overline{c'w'}\right)}{\partial z} \qquad (29)$$

where c' = fluctuation of sediment concentration; c = local time-averaged sediment concentration, u, v and w are the streamwise, vertical and lateral time-averaged velocities; u', v' and w' are the velocity fluctuations in y and z directions, respectively.

In equilibrium conditions, time averaging of Eq. (29) gives:

$$\frac{\partial\left(cv + \overline{c'v'} - c\omega\right)}{\partial y} = 0 \qquad (30)$$

The integration of Eq. (30) with respect to y yields the following equation

$$cv + \overline{c'v'} - c\omega = 0 \qquad (31)$$

If the eddy viscosity is used and Rouse number in Rouse's law has the following form:

$$Z(Y) = \frac{\omega(1 - Y)}{\kappa u_*} \qquad (32)$$

$$\frac{Z(Y)}{Z(0)} = 1 - Y \qquad (33)$$

Similar to the Shields number, many researchers also found that the measured Z is different from the calculated Z. *van Rijn* [10] and *Van de Graaff* [38] attribute this invalidity to sediment characteristics like size or streamwise flow strength, Eq. (33) indicates that if the vertical velocity exists, it also leads to the invalidity of Rouse number in practice.

Rosea and Thorneb [39] observed the Rouse number by measuring suspended sediment concentration profiles in the river Taw estuary, UK, where the flow is dominated by strong rectilinear, turbulent tidal currents. Their measurement was focused on the rising (flood) tide for a period of 3 hours. The measured $Z(Y)/Z(0)$ is shown in **Figure 5**, at the at the starting point the minimum vertical velocity Y can be expected, and $Z(Y)/Z(0) \approx 1$ is observed, in the process, the streamwise velocity or shear velocity changed in a range of $\pm 20\%$, but the observed $Z(Y)/Z(0)$

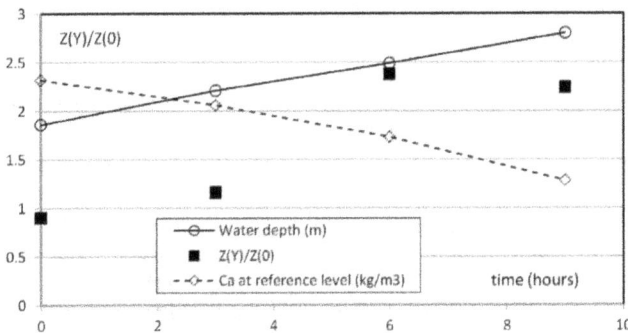

Figure 5.
Measured Z(Y)/Z(0) in a rising tidal flow by Rose and Thome, at the starting point the streamwise velocity was the highest, minimum vertical velocity Y can be inferred, and Z(Y)/Z(0) ≈ 1 is observed and all data points show Z(Y)/Z(0) >1 in the rising tidal flow.

Profile	1	2	3	4	5	6	7	8	1	2	3	5
Tide		ebb							Flood			
Rouse number	0.55	0.45	0.45	0.50	0.27	0.35	0.48	0.50	0.68	0.95	0.7	1.2
Average		0.44							0.70			

Table 1.
Measured rouse numbers ($\omega/\kappa u_*$) in flood-tide and ebb tide by AL-Ragum [3].

increased 150% and all data points shows that $Z(Y)/Z(0) >1$ in the rising tidal flow, this is agreed with Eq. (33), i.e., accelerating flows generate an downward velocity or negative Y that constrains sediment transport. This also can be seen from the measured sediment concentration Ca at the reference level near the sea bed, the decreasing Ca implies that the downward velocity makes the particles "heavier" to move, consequently Ca is reduced to 44.8% of its original value.

If $Z(Y)/Z(0)$ in flood tide is compared with its values observed during ebb-tidal, Eq. (33) clearly indicates that the ebb-tide will have a lower value. This is in agreement with *Al-Ragum's* [40] observation as shown in **Table 1**. The data were collected from the Biscay Bay near Spain and France border. "The Rouse parameter varied with the tide, and the values were higher on the flood-tide than on the ebb-tide" as claimed by the author. The average Rouse parameter during flood tide is about 0.7, but it is reduced to 0.44 during the ebb tide. The flood tide generates 60% higher Rouse number relative to that during the ebb-tide.

6. Discussion on vertical velocity induced by unsteadiness and its effects

6.1 Unsteadiness parameter

For sediment transport by either flood waves in rivers or tidal waves in the sea, the unsteadiness plays a significantly role for sediment transport. The equations developed from steady flow may be invalid in unsteady flows. Some researchers like *Graf and Suszka* [41] found that the measured sediment transport rate in an unsteady flow is always larger than these equations' predictions. An unsteadiness parameter was proposed by them:

$$P_1 = \frac{h_p - h_1}{t_d u_*} \tag{34}$$

where t_d is the duration of a hydrograph, h_1 is the initial or baseflow depth, h_p is the peak flow depth of the hydrograph.

It is interesting to note that $(h_p - h_1)/t_d$ is actually the averaged vertical velocity V. Eq. (34) can be understood as the ratio of vertical velocity to the shear velocity, similar to Y in Eq. (19). The unsteadiness parameter P_1 is useful for the prediction of time average sediment transport rate, but it cannot be used to explain the measured instantaneous rate g_t or concentration C that depends on the instantaneous vertical velocity, thus Eq. (19) may have a wider application. Compared with Eq. (34), Eq. (19) is simple and direct, the difficult parameter u_* is replaced with the sediment settling velocity ω that is independent of flow characteristics, and the instantaneous vertical velocity V can easily explain the observed phenomena in unsteady flows.

In fact, the unsteadiness parameter S_1 in Eq. (34) can be written in its alternative form:

$$S_1 = \frac{V}{(\rho_s/\rho - 1)ga/U} \tag{35}$$

where a is the wave amplitude similar to h_p-h_1 in Eq. (34) and the vertical velocity $V = a\sigma/(2\pi)$, thus Eq. (35) shows that S_1 is similar to Y. *Sleath* [18] also proposed another parameter to express sediment transport by waves, i.e., $\omega/\sigma\delta$, and δ is the maximum thickness of the mobile layer, which can be read as $1/Y$ if $V_b = \sigma\delta$ is assumed.

Figures 4 and **5** show the influence of unsteadiness on sediment transport in tidal flows. For flood waves in a river, *Lee et al.* [15] measured the transport rate over a series of triangular hydrographs. Their experimental results show the existence of phase lag between the peak discharge and peak sediment rate g_t, which lag is very long and about 6–15% of the flow hydrograph duration. **Figure 6** shows the hydrograph and measured sediment discharged by Lee et al. [15], it shows that the highest sediment transport rate appears in the falling stage when the shear stress is much less than that in the rising stage. This phase-lag phenomenon cannot be explained by those shown in Eqs. (1)–(7). It is interesting to note that there are two g_t peaks in **Figure 6**, the mechanisms may be totally different, the former in the rising limb is likely generated by very high τ in the rising limb, but the upward velocity probably dominates the second peak where the shear stress is very small.

It should be mentioned that the peak sediment discharge in the rising limb is not always discernible as shown in **Figure 6**. For example, *Qi et al.* [42] reported that in Yellow River, artificial flood waves have been used to flush sediment in lower course of Yellow River by releasing water from its Xiaolangdi reservoir. As shown in **Figure 7**, the rising limb did not increase sediment concentration much, but the falling stage generated very high sediment concentration. From their experience, to enhance the flush efficiency, the duration of rising limb should be short as its rising flow does not increase g_t or C too much.

To interpret the results in **Figures 6** and **7**, the conceptual mathematical model in Eq. (27) may be useful as it covers the parameters in streamwise and vertical directions. Eq. (27) precisely suggests that the upward velocity may be responsible for the widely observed "phase lag" in sediment transport in rivers.

6.2 Mechanism of bedform formation

The formation of ripples and dunes over a flat mobile bed is an amazing phenomenon, and has attracted many investigations. All previous equations of

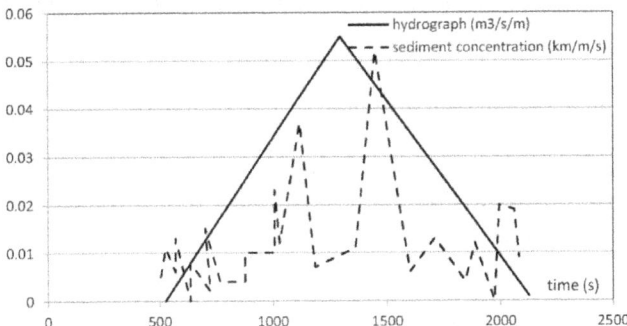

Figure 6.
Sediment transport rate and flood hydrograph measured by Lee et al. [23].

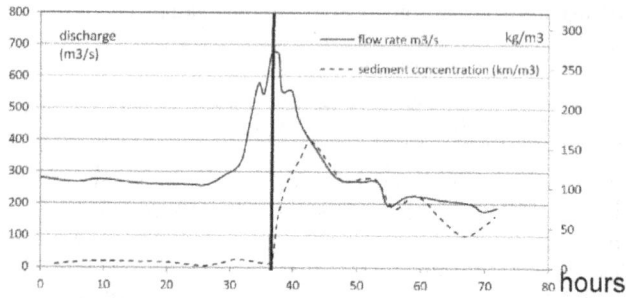

Figure 7.
Measured sediment concentration over a hydrograph at Huayuankou, Yellow River from July 4–6, 2010 by Qi et al. [31].

sediment transport (e.g., Eqs. (1)–(7) and Eqs. (34) and (35) fail to explain how the bedforms are formed, because these equations only use the streamwise parameters (U, τ etc.) that are constant in every cross section from upstream to downstream if the flow is steady and uniform, thus the sediment discharge in every cross section is same and no local erosion occurs, so none of them can successfully explain the formation of ripples and dunes.

However, Eqs. (14) and (22) may provide a possible explanation for the discontinuity of sediment transport from upstream to downstream. It is well known that turbulence in a steady and uniform flow is dominated by complex, multiscaled, quasi-random and organized eddies that possess both spatial and temporal coherence [43]. The velocity fluctuations are also governed by the continuity equation with the following form:

$$v' = -\int \frac{\partial u'}{\partial x} dy \qquad (36)$$

The coherent events can be broadly divided into ejections ($v' > 0$ or decelerating) and sweeps ($v' < 0$ or accelerating), both of them are always alternated in space and time.

To help conceive the formation of bedforms, a flow region in **Figure 8** is divided simply into three zones, A, B and C during a short period. If the flow region B is dominated by the ejection event (denoted by "+" in **Figure 8** for upward vertical velocity), severe erosion should be observable in Zone B as Eq. (27) and **Figure 4** indicate that the upward velocity significantly promotes the sediment discharge. On the other hand, Zones A and C are dominated by the downward velocity (or negative "-" velocity), and Eq. (27) and **Figure 4** predict that the sediment carrying capacity is weaker if the vertical velocity is negative, therefore the sediment from zone B has to deposit at Zone C. It can be seen that the vertical velocity and its alternation in direction in space play a key role for the formation of dunes and ripples. The discontinuity of sediment-laden capacity along the flow direction is uneven, this triggers the formation of bedforms, once some scouring holes are formed over a flat mobile bed, erosion in these areas most likely would continue till the equilibrium condition is reached.

Alternatively, we can consider a simple model that all particles in **Figure 8** possess higher apparent density in zone A and C like iron particles (represented by dark solid circles in **Figure 8**), but the particles in zone B have lighter density (like plastic particles). All particles in zone A, B and C have the same diameter. It is understandable that a scour hole will be formed in zone B, and deposition will occur

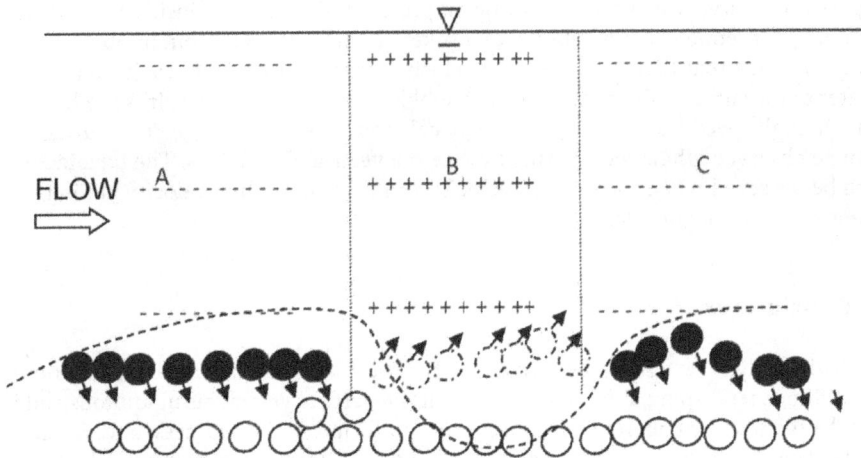

Figure 8.
Relationship between the alternative vertical velocity and bedform formation, where "+" sign denotes upward velocity in region B and "-" is the downward velocity in region A and C. The dotted vertical lines denote the flow region division lines, the open circles denote the sediment particles, the solid circles denote that particles' density "becomes heavier", and dashed circles denote the "lightweight sediment", the open circles are normal sediment particles.

at C even though the U and τ remains constant in zones A to C. In other words, it can be seen that the vertical velocity and its spatial alternation play a key role for the formation of bedform. The simple model shown in **Figure 8** explains the formation of a scour hole on a flat plane that triggers the formation of bedforms. This mechanism can be extended to dune formation in deserts where the horizontal wind generates sediment transport in horizontal direct, and vertical motions yields the bedforms. The wind is accelerating along the upwind side of a dune, thus its surface is smooth, and the decelerating wind after the peak generates upward velocity, thus small holes are formed in the lee side.

By comparing the mechanism of phase lag and bedform formation discussed above, one may find that the vertical velocity is responsible for both phase lag phenomenon and bedform formation. The temporal alternation of upward and downward velocities generates the phase lag phenomenon, whilst its spatial alternation yields the bedforms. Generally speaking, we can see that the phenomena of sediment transport can be categorized into streamwise and vertical motions dominated events. Sediment transport should be expressed using variables in streamwise and vertical directions jointly.

6.3 Unifying mechanism of wave formation and breaking waves

Generally, all interfaces on solid–liquid, liquid–liquid, liquid-gaseous phases exist waves if there exists alterative vertical motions as shown in **Figure 8**, otherwise no waves can be observed no matter how high the velocity is if the flow is laminar. Likewise, the ocean waves between water and air are not caused by the shear stress or wind velocity on the sea surface, but the air pressure oscillation whose period should be identical to the ocean waves. In other words, turbulence is the cause of ocean waves. In summer, the heated sea surface generates an upward motion, consequently typhoons, cyclones and hurricanes can be observed. In winter, the downward cold air yields a relatively calm surface.

The existence of upward velocity can be inferred from numerous small bubbles when waves are broken. The soluble gas or air near in a lower lever like the seabed (high pressure) can be transferred to the surface (gauge pressure = 0) by the upward velocity, which causes significant pressure difference of inside and outside bubbles, consequently the bubbles are broken. In other words, from bubbles one may conclude that there is an upward velocity to transfer the bubbles from deep water to the surface, this is also true for bubbles in hydraulic jumps. It is predictable that in high speed flow, cavitation (i.e., local scour over a metal/concrete surface) can be observed when decelerating flow or the vertical flow exists. The liquefaction can be observed when the seepage velocity and particle settling velocity are in the same order of magnitude.

7. Conclusions

This study investigates the influence of vertical velocity induced unsteady flows on sediment transport. It is well-known that the vertical velocity is ubiquitous and it can be induced by coherent structures, non-uniformity, unsteadiness, and so on. This paper just discusses the simplest cases, i.e., the presence of vertical velocity does not significantly alter the streamwise parameters like velocity U or discharge q, in which the rising limb or accelerating flow generates a downward velocity, but the falling limb or decelerating flow induces an upward velocity. A conceptual mathematical model is developed to account for the vertical velocity's influence on particles' critical shear stress, sediment discharge and suspension. It is found that the model can provide a qualitatively explanation to some observed phenomena. Based on this investigation, the following conclusions can be drawn:

1. The upward velocity enhances sediment mobility and downward velocity increases its stability. Mathematically the behavior of sediment transport subject to a vertical motion can be equivalently treated by the variation of apparent density. Particles become "heavier" when they experience the downward flows, this reduces the sediment transport rate. But particles become "lighter" in flows with upward velocity where the sediment discharge is increased significantly. The obtained new equation for sediment transport's apparent density is used to explain sediment transport in unsteady flows.

2. The application of Shields diagram, equations of sediment discharge and Rouse equation developed from steady flows could be extended to unsteady flows if the vertical parameter Y $(= V/\omega)$ is included. The conceptual model shows that sediment is easily be transported when $Y > 0$, but difficult to move when $Y < 0$, same for the transport rate g_t and Rouse number Z. The developed equations provide reasonably good agreement with the measured data. The condition for liquefaction can be expressed by $Y = 1$.

3. The mathematical model may also provide a tool to understand many odd phenomena in sediment transport like the phase lag phenomenon and bedform formation. Both are widely reported and discussed, this is the first trail to give the similarities between these two phenomena. The research shows that the temporal variation of vertical velocity results in the phase lag, and its spatial variation leads to the bedform formation.

4. In the literature, the vertical velocity is generally ignored in the measurement, which leads to that the conclusions listed above rest on the inferences of vertical velocity, not its measured values and direction. In future, systematical experiments are needed to investigate its role in order to validate the conceptual model.

Notations

a = *wave amplitude*

a_0 = *wave-orbital acceleration*

C = averaged sediment concentration by volume

c = time-average point concentration

c' = fluctuation of sediment concentration

c_f = friction factor

d = particle diameter

d_* = *dimensionless particle size*

E = energy dissipated on skin friction (i.e., $\tau u_*'$)

g = gravitational acceleration

g_t = sediment discharge

h = water depth

H = wave height

h_1 = depth of baseflow

h_p = peak flow depth of the hydrograph

i = hydraulic gradient;

k = constant

k_1 = factor

k_0 = wave number

L = wave length

m_1 = coefficient

n = the number of zero-downcrossing waves in the burst

P = pressure

q = discharge per unit width

S = energy slope

S_1 = Sleath number

t = time

T = transport parameter defined by van Rijn

T_{ZDC} = the period for the events

t_d = duration of a hydrograph

U = mean velocity

u_* = shear velocity

u_{*c} = critical shear velocity

u_*' = shear velocity related to grain friction

u, v, w = time-averaged velocity in the streamwise, vertical direction and spanwise directions

u', v' and w' = velocity fluctuation in streamwise and vertical and spanwise directions

U_{ZDC+} = the maximum current excursion in the positive direction

U_{ZDC-} = the maximum current excursion in the negative direction

V = vertical velocity

U_b = wave-orbital speed at the bed

V_b = vertical velocity at the bed

x = streamwise direction

y = vertical direction

z = lateral direction

$Y = V_b/\omega$

Z = Rouse number

β = coefficient

δ = maximum thickness of the mobile layer

ν = kinematic viscosity

ρ = fluid density
ρ_s = density of sediment
ρ_s' = sediment apparent density
σ = angular frequency of waves
τ = boundary shear stress
τ_* = Shields number
τ_*' = Shields number with vertical velocity
τ_c = critical shear stress
τ_c' = critical shear stress with vertical velocity
ω = particle fall velocity
subscribes b and t = bed load and the total load

Author details

Shu-Qing Yang
School of Civil, Mining and Environmental Engineering, Faculty of Engineering,
University of Wollongong, Australia

*Address all correspondence to: shuqing@uow.edu.au

IntechOpen

References

[1] Einstein HA. Formulas for the transportation of bed load. Trans. Soc. Civ. Engrg. 1942;**107**:561-597

[2] Meyer-Peter E, Muller R. "Formula for Bed Load Transport." Proc., 2nd Meeting, Vol. Stockholm: IAHR; 1948. p. 6

[3] Yalin MS. Mechanics of Sediment Transport. Oxford, U.K.: Pergamon; 1977

[4] Engelund F, Hansen E. A Monograph on Sediment Transport in Alluvial Streams. Copenhagen, Denmark: Teknisk Forlag; 1972

[5] Ackers P, White WR. Sediment transport: New approach and analysis. J. Hydr. Div. 1973;**99**(11):2041-2060

[6] Velikanov, M. A. (1954). "Gravitational theory for sediment transport." J.of science of the Soviet Union, Geophysics, Vol. 4, (in Russian).

[7] Chien N and Wan Z. (1999), Mechanics of Sediment Transport, ASCE Press.

[8] Bagnold RA. An Approach to the Sediment Transport Problem from General Physics." Geol. Survey Professional Paper No. 422-I, U.S. Washington, D.C.: Government Printing Office; 1966

[9] Yang, C. T. (1973). "Incipient motion and sediment transport." J. Hydraul. Div., Am. Soc. Civil Engineering, 99 (10), 1679–1704.

[10] van Rijn LC. Sediment transport, Part II: Suspended load transport. J. Hydr. Engrg., ASCE. 1984;**110**(11):1613-1641

[11] Yang SQ, Tan SK. Flow resistance over mobile bed in an open-channel flow. Journal of Hydraulic Engineering, ASCE. 2008;**134**(7):937-947

[12] Yang SQ, Kim I-S, Koh DS, Song YC. Interaction of Streamwise and Wall-Normal Velocities in Combined Wave-Current Motion. China Ocean Eng. 2005;**19**(4):557-570

[13] Yang SQ, Koh SC, Kim IS, Song YC. Sediment transport capacity-an improved Bagnold formula. International Journal of Sediment Research. 2007;**22**(1):27-38

[14] Liu X.X and Chiew Y.M., (2012), "Effect of Seepage on Initiation of Cohesionless Sediment Transport". Acta Geophysica 60(6), 1778–1796. DOI: 10.2478/s11600-012-0043-7.

[15] Lee KT, Liu YL, Cheng KH. Experimental investigation of bedload transport processes under unsteady flow conditions. Hydrological Processes. 2004;**18**:2439-2454. DOI: 10.1002/hyp.1473.

[16] Nelson JM, Shreve RL, McLean SR, Drake TG. Role of near-bed turbulence structure in bed load transport and bed form mechanics. Water Resources Rese. 1995;**31**(8):2071-2086

[17] Tabarestani, M.K. and Zarrati, A.R. (2015). Sediment transport during flood event: a review, Int. J. Environ. Sci. Technol. 12:775–788, DOI 10.1007/s13762-014-0689-6

[18] Sleath JFA. Conditions for plug formation in oscillatory flow. Continental Shelf Research. 1999;**19**(13):1643-1664

[19] Francalanci S, Parker G, Solari L. Effect of seepage-induced nonhydrostatic pressure distribution on bed-load transport and bed Morphodynamics. J. Hydr. Eng., ASCE. 2008;**134**(4):378-389

[20] Cheng NS, Chiew YM. Incipient sediment motion with upward seepage.

J. Hydraul. Res. IAHR. 1999;**37**(5): 665-681

[21] Chen MS, Wartel1 S, Eck BV, Maldegem DV. "Suspended matter in the Scheldt estuary". Hydrobiologia. P. Meire & S. van Damme (eds), Ecological Structures and Functions in the Scheldt Estuary: from Past to Future. 540. 2005:79-104

[22] Wren DG, Kuhnle RA, Wilson CG. Measurements of the relationship between turbulence and sediment in suspension over mobile sand dunes in a laboratory flume. J. Geophy. Res. 2007; **112**. DOI: F03009, 10.1029/ 2006JF000683

[23] Cellino M, Lemmin U. Influence of coherent flow structures on the dynamics of suspended sediment transport in Open-Channel flow. Journal of Hydraulic Engineering. 2004;**130** (11):1077-1088

[24] Song T, Graf WH. Velocity and trbukence distribution in unsteady open-channel flows. J. Hydr. Eng., ASCE. 1996;**122**(3):141-154

[25] Leng X and Chanson (2016), Unsteady Turbulent Velocity Profiling in Open Channel Flows and Tidal Bores using a Vectrino Profiler. HYDRAULIC MODEL REPORT No. CH101/15, ISBN 978–1–74272-145-3, The University of Queensland, School of Civil Engineering.

[26] Schlichting H. Boundary-Layer Theory. 7th ed. New York: McGraw-Hill Book Company; 1979

[27] Kemp, P. H. and Simons, R. R.ʻ (1982). The interaction between waves and a turbulent current:waves propagating with the currentʻJ. Fluid Mech., 116ʻ227 ~ 250.

[28] Kemp, P. H. and Simons, R. R.ʻ (1983). The interaction of waves and turbulent current:waves propagating

against the currentʻJ. Fluid Mech.ʻ130ʻ 73 ~ 89.

[29] Yang SQ, Tan SK, Wang XK. Mechanism of secondary currents in open channel flows, J. Geophysical Research. 2012;**117**:F04014. DOI: 10.1029/2012JF002510

[30] Yang SQ, Chow AT. Turbulence structures in non-uniform flows. Advances in Water Resources. 2008;**31**: 1344-1351

[31] Fredsoe, J. and Deigaard, R. (1992). Mechanics of Coastal Sediment Transport. Advanced Series on Ocean Eng., Vol. 3, World Scientific, Singapore.

[32] Nielsen P. Coastal Bottom Boundary Layers and Sediment Transport, Advanced Series on Ocean Eng. Vol. 4. Singapore: World Scientific; 1992

[33] Green MO, MacDonald IT. Processes driving estuary infilling by marine sands on an embayed coast. Marine Geology. 2001;**178**(1/4):11-37

[34] Green MO, Coco G. Review of wave-driven sediment resuspension and transport in estuaries. Reviews of Geophysics. 2014;**52**:77-117. DOI: 10.1002/2013RG000437.

[35] Yang SQ. Formula for sediment transport in rivers, estuaries and coastal waters. Journal of Hydraulic Engineering, ASCE. 2005;**131**(11):968- 979

[36] Green MO. Very small waves and associated sediment resuspension on an estuarine intertidal flat. Estuarine, Coastal and Shelf Science. 2011;**93**(4): 449-459

[37] Yang SQ. Turbulent transfer mechanism in sediment-laden flow. Journal of Geophysical Research, AGU. 2007;**112**. DOI: F01005, 10.1029/ 2005JF000452

[38] Van de Graaff J. Sediment
Concentration Due to Wave Action.
Diss. Delft Univ. of Technology; 1988

[39] Rosea CP, Thorneb PD.
Measurements of suspended sediment
transport parameters in a tidal estuary.
Continental Shelf Research. 2001;21:
1551-1575

[40] AL-Ragum AN. An evaluation of
the rouse theory for the suspension of
sand in a tidal inlet. In: PhD Thesis. UK:
Univ. of Southampton; 2015

[41] Graf WH, Suszka L. Sediment
transport in steep channels. J.
Hydroscience., Hydr. Eng. 1985;5(1):
11-26

[42] Qi P, Qu SJ, Sun ZY. Suggestion on
optimizing flow and sediment
regulations as the operational methods
of Xiaolanfdi reservoir. Yellow River.
2012;34(1):5-9

[43] Adrian RJ. Hairpin vortex
organization in wall turbulence. Physics
of Fluids. 2007;19:041601

Physics of Cohesive Sediment Flocculation and Transport: State-of-the-Art Experimental and Numerical Techniques

Bernhard Vowinckel, Kunpeng Zhao, Leiping Ye,
Andrew J. Manning, Tian-Jian Hsu, Eckart Meiburg
and Bofeng Bai

Abstract

Due to climate change, sea level rise and anthropogenic development, coastal communities have been facing increasing threats from flooding, land loss, and deterioration of water quality, to name just a few. Most of these pressing problems are directly or indirectly associated with the transport of cohesive fine-grained sediments that form porous aggregates of particles, called flocs. Through their complex structures, flocs are vehicles for the transport of organic carbon, nutrients, and contaminants. Most coastal/estuarine models neglect the flocculation process, which poses a considerable limitation of their predictive capability. We describe a set of experimental and numerical tools that represent the state-of-the-art and can, if combined properly, yield answers to many of the aforementioned issues. In particular, we cover floc measurement techniques and strategies for grain-resolving simulations that can be used as an accurate and efficient means to generate highly-resolved data under idealized conditions. These data feed into continuum models in terms of population balance equations to describe the temporal evolution of flocs. The combined approach allows for a comprehensive investigation across the scales of individual particles, turbulence and the bottom boundary layer to gain a better understanding of the fundamental dynamics of flocculation and their impact on fine-grained sediment transport.

Keywords: cohesive sediment, floc measurement, particle-resolved direct numerical simulation, continuum model, population balance equation

1. Introduction

Cohesive sediment transport is a tightly coupled system driven by hydrodynamic forcing, resuspension, deposition, and flocculation. Turbulence intensity in the bottom boundary layers and the water column mixed layers controlled by currents (e.g., river outflows, tidal currents), surface waves, and estuarine circulations can vary by several orders of magnitude in terms of turbulent dissipation rate. Hence, the turbulent shear rate defined as $G = \sqrt{\varepsilon/\nu}$ is the most important parameter for flocculation models. In the estuarine tidal boundary layers, the turbulent dissipation rate is around $\mathcal{O}(10^{-6} \sim 10^{-4})m^2/s^3$ ($G = 1 \sim 10s^{-1}$; e.g., [1, 2]). Near estuarine fronts with shear instabilities, turbulent dissipation rates can increase up to $\mathcal{O}(10^{-3})m^2/s^3$ ($G = 30s^{-1}$; [3]). During storm condition when wave motion become dominant, turbulent dissipation rates in the thin wave bottom boundary layer can exceed $\mathcal{O}(10^{-3})$ ($G = 30 \sim 100s^{-1}$, [4, 5]). Under breaking waves in the upper ocean or in the surf zone, the turbulent dissipation rate can be as high as $\mathcal{O}(10^{-3} \sim 10^{-2})$ ($G = 100 \sim 200s^{-1}$, [6, 7]).

Computer simulation models are commonly the chosen tools that coastal managers use to predict sediment transport rates. In these continuum models, the fluid motion is computed via the continuity equation and the Navier-Stokes equations or spatially and temporally averaged variants thereof, whereas the sediment is represented as a concentration field [8]. This approach allows to simulate large spatial scales covering entire estuaries, because the governing equations are solved on the same Eulerian grid. However, these types of models require closures to account for the unresolved physics of the sediment dynamics, in particular vertical sedimentary distributions [9, 10] and mass fluxes. The latter is the product of the concentration and the settling velocity. Manning and Bass [11] found that mass settling fluxes can vary over four or five orders of magnitude during a tidal cycle in mesotidal and macrotidal estuaries; therefore, a realistic representation of flux variations is crucial to an accurate depositional model.

The specification of the flocculation term within numerical models depends on the sophistication of the model structure. Until recently, even the conceptual relationship between floc size, suspended particulate matter (SPM) concentration and turbulent shear stress proposed by Dyer [12] (see **Figure 1a**) remained largely unproven. Hence, much more work needs to be done in order to arrive at robust simulation tools with predictive capacity. The present contribution, therefore, provides a review on the state-of-the-art of floc measurements in both the field and laboratory. In addition, we review a newly emerging technique of particle-resolved simulations that can provide a promising alternative avenue to generate data of

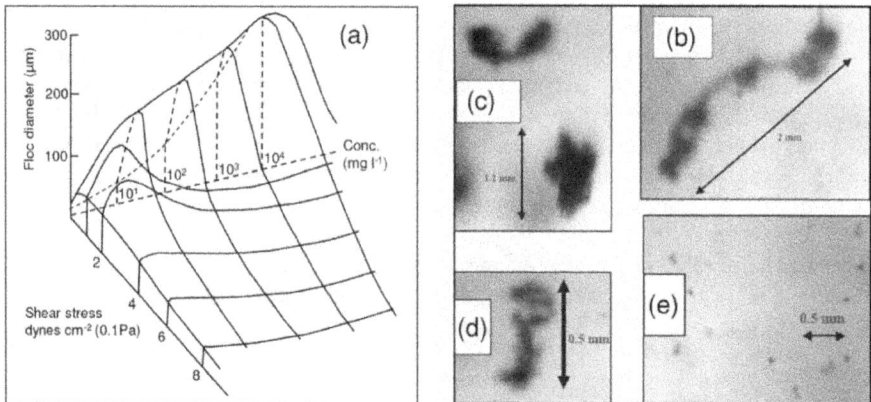

Figure 1.
An illustration of the dependence of floc size on shear stress and SPM concentration is given. (a) A conceptual diagram showing the relationship between floc modal diameter, suspended sediment concentration and shear stress (from [12]). Subplots (b–e): illustrative examples of real estuarine floc images showing ambient shear stress, SPM concentration and settling velocity; (b) shear stress 0.3 N m^{-2}, SPM concentration 3.5 g l^{-1}, and settling velocity 5 mm s^{-1}; (c) shear stress 0.3 N m^{-2}, SPM concentration 3.5 g l^{-1}, and settling velocity 8 mm s^{-1}; (d) shear stress 0.45 N m^{-2}, SPM concentration 0.25 g l^{-1}, and settling velocity 1.8 mm s^{-1}; (e) shear stress 0.45 N m^{-2}, SPM concentration 0.25 g l^{-1}, and settling velocity 0.2 mm s^{-1}. Illustrations (b–e) are modified from Manning.

small-scale sediment dynamics to derive the missing constitutive equations for continuum models. Finally, we provide a summary of current techniques that are used in continuum models to account for cohesive sediment dynamics, where we explicitly point to the components that deserve more research in the future.

2. Assessment of temporal floc evolution

2.1 Floc measurements

2.1.1 Direct floc size measurements

The presence of large estuarine macroflocs was initially observed in situ using underwater photography [13]. However, floc breakage occurs during sampling in response to the additional shear created by the instrumentation [14]. To overcome this problem, less-invasive techniques for measuring floc properties in situ have been developed. Usually, these can be divided into devices that solely measure floc size (D) e.g. Lasentec (**Figure 2a**) [15], LISST [16], LISST-Holo [17], and InSiPid [18]; and those devices that can provide measurements both of floc size and settling velocity (W_s) e.g. VIS [19], HR Wallingford video camera system [20], VIL [21], INSSEV: IN-Situ SEttling Velocity instrument [22, 23] (**Figure 2b–c**), LabSFLOC—

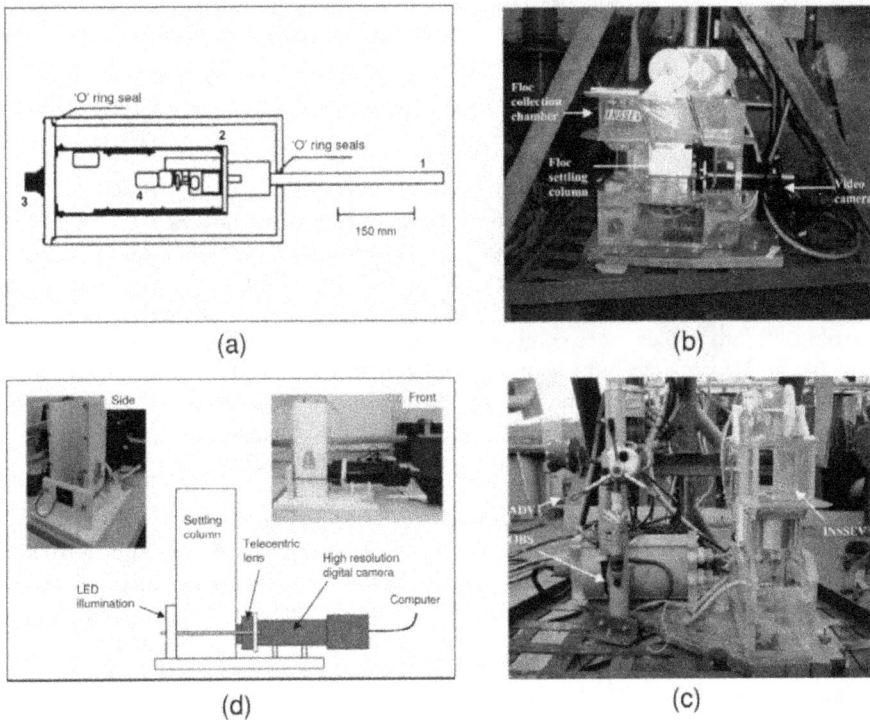

(a)   (b)

(d)   (c)

Figure 2.
Some examples are given of floc measuring instrumentation. (a) Schematic of the adapted Lasentec par-tec 100 probe unit showing: 1. The light guide, scanning mechanism and focusing lens, 2. The PVC cylinder, 3. Watertight cable termination, and 4. Electronic circuitry and power supplies (from [15]). (b, c) the INSSEV instrument (from [24]); (b) side view of INSSEV mounted on a metal deployment frame; (c) front view of INSSEV (right side of image), together with optical backscatter (OBS) sensors and an acoustic Doppler velocimeter (ADV) positioned on a vertical pole (left side of image). The ADV provides high frequency turbulence data that can be directly related to the floc populations. (d) Views of a LabSFLOC-2 together with a schematic illustration of the instrument after [25]).

Laboratory Spectral Flocculation Characteristics—instrument [25–27], INSSEV-LF [28], and PICS—Particle Imaging Camera System [29].

The strength of video-based floc measurements is that they minimize the number of assumptions used during the data processing and interpretation stages. Devices that only measure the size component require additional gross and often incorrect assumptions regarding the relationship between settling velocity, floc size, and floc density. The settling velocity of a floc is a function of both its size and effective density, and both of these floc components can display variations spanning three to four orders of magnitude within any one floc population [30–32].

Of note, the LabSFLOC suite of high-resolution, low intrusive, underwater video camera systems for the past 20+ years have been regarded internationally as a benchmark device for the sampling and dynamical simultaneous measurement of sizes and settling velocities for entire populations of flocs and a range of bio-sedimentary particles, and this enables individual floc effective density values to be determined by applying a modified Stokes Law [33]. This also enables the calculation of additional floc properties including: structural composition (porosity, fractal dimension), shape, sedimentary mass flux, and floc population mass-balancing, all from within a wide range of aquatic environments. In conclusion, selection of the most appropriate instrumentation is paramount when attempting to parameterise flocculated cohesive sediments. Manning et al. [34] provide a detailed review of many of these floc measuring systems.

2.1.2 Parameterizing floc data

To aid the interpretation of floc characteristics and for their inclusion in sediment transport models, each floc population can be segregated into various subgroupings according to floc size. Sample-mean floc values can be computed (i.e., a single value per floc population) to show generalized floc property trends.

Dyer et al. [35] reported that a single mean or median settling velocity did not adequately represent an entire floc spectrum, especially in considerations of a flux to the bed. Dyer et al. [35] recommended that the best approach for accurately representing the settling characteristics of a floc population was to split a floc distribution into two or more components, each with their own mean settling velocity. Both Eisma [13] and Manning [32] concur with this finding by suggesting that a more realistic and accurate generalization of floc behavior can be derived from the macrofloc and microfloc fractions. These two floc fractions form part of Krone's [36] classic order-of-aggregation theory and produce two floc property values per floc population.

Macroflocs are large (typically $D > 160$ μm [32]) highly porous (typically $> 90\%$) and often fragile (and difficult to sample), fast-settling aggregates (see **Figure 1b–d**), and typically close to the size of the turbulent Kolmogorov microscale [37, 38]. They are recognized as the most important subgroup of flocs because their fast-settling velocities tend to have the strongest influence on the mass settling flux [39]. Macroflocs are progressively broken down as they pass through regions of higher turbulent shear stress and reduced again to their component microfloc substructures [40]; they rapidly attain equilibrium with the local turbulent environment.

The smaller microflocs (typically $D < 160$ μm [32]; see **Figure 1e**) are considered to be the building blocks from which the macroflocs are comprised. Many field studies [22, 41–43] have shown that the microfloc class of aggregates tend to display a much wider range of effective densities and settling velocities than the macrofloc fraction. Microflocs are much more resistant to breakup by turbulent shear; they tend to have slower settling velocities but exhibit a much wider range of effective densities than the larger macroflocs.

In terms of flocculation kinetics [44], the macroflocs tend to control the fate of purely muddy sediments in an estuary [45]; this is because the smaller microflocs generally settle at less than 1 mm s^{-1}, whereas macroflocs settle in the 1–15 mm s^{-1} range, enabling them to deposit to the bed [46]. However, when flocculation of mixed sediments occurs, the microflocs can potentially demonstrate settling velocities comparable to those of the macroflocs [28, 47].

2.1.3 Floc data example

In order to illustrate the spectral variability of floc properties, each floc population can be divided into various size bands. The band divisions can be chosen to best fit the data collected. LabSFLOC data for a mud sample from the Medway Estuary, UK, provide a graphical illustration of size banding and the increasing settling velocities associated with larger flocs (**Figure 3**). Twelve size bands (SB) have been used to represent the Medway floc data, with SB1 representing microflocs less than

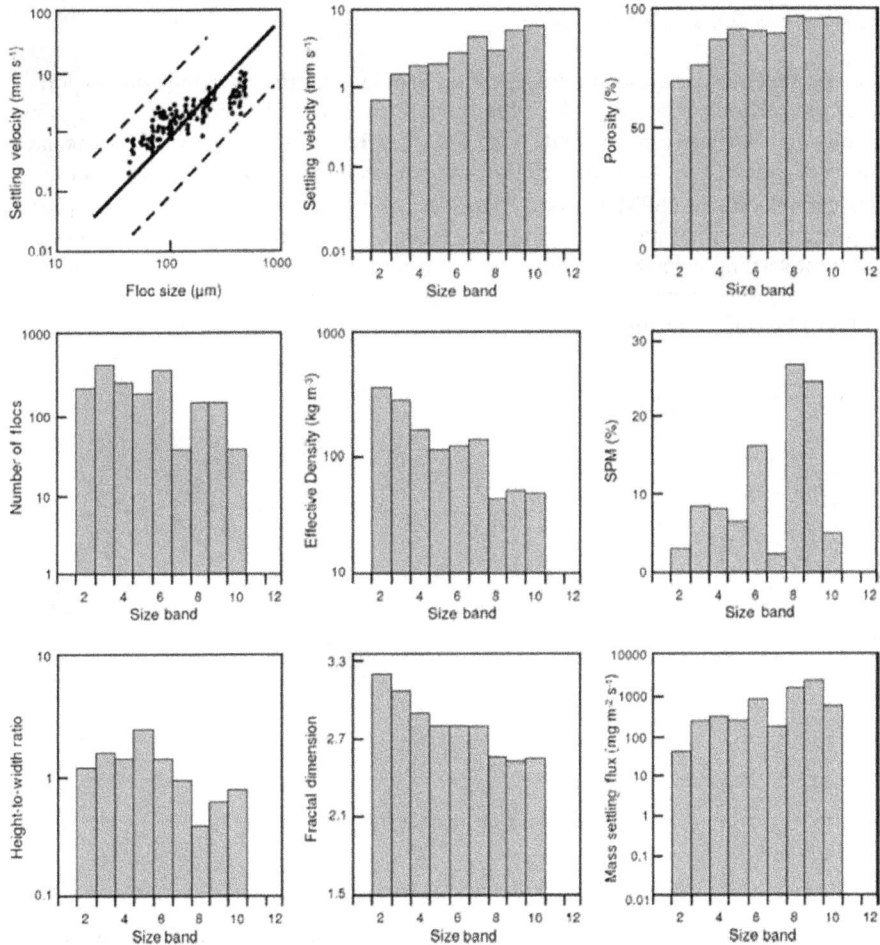

Figure 3.
An example of various size-banded floc properties for a LabSFLOC SPM sample is given. The complete population of size versus settling velocity data is illustrated in the top-left panel. These data apply to Medway estuary mud slurry (1.6 g l^{-1}) that has been sheared in the Southampton oceanography Centre (UK) mini-flume at a shear stress of 0.37 nm^{-2} (data from results in [43]).

40 μm in size, whilst SB12 is representative of macroflocs greater than 640 μm in diameter; SB2 to SB6 range from 40 to 240 μm in five steps, each of 40 μm, and SB7 to SB11 range from 240 to 640 μmm in five steps of 80 μm.

2.1.4 Floc settling modeling approaches

2.1.4.1 Constant settling velocity

Specification of the flocculation term within numerical models depends on the sophistication of the model. The simplest parameterisation is a single floc settling velocity value that remains constant in both time and space (one coefficient). These fixed settling values are usually in the range of 0.5–5 mm s^{-1} and typically are selected on an arbitrary basis and sometimes used as a tuning parameter to match predicted erosion and deposition patterns to observations for an undisturbed estuary.

2.1.4.2 Power law settling velocity

The next step is to use gravimetric data provided by field settling-tube experiments to relate floc settling velocity to the instantaneous SPM concentration, using a power law with two coefficients (e.g., [48]; see **Figure 4a**). Empirical results have shown a generally exponential relationship between the mean, or median, floc settling velocity and SPM concentrations for concentrations <10 g l^{-1}. This approach sometimes includes hindered settling (see **Figure 4b**). However, both the constant settling velocity and the power law parameterisation techniques do not include the important and influential effects of turbulence such as floc breakup induced by turbulent shear.

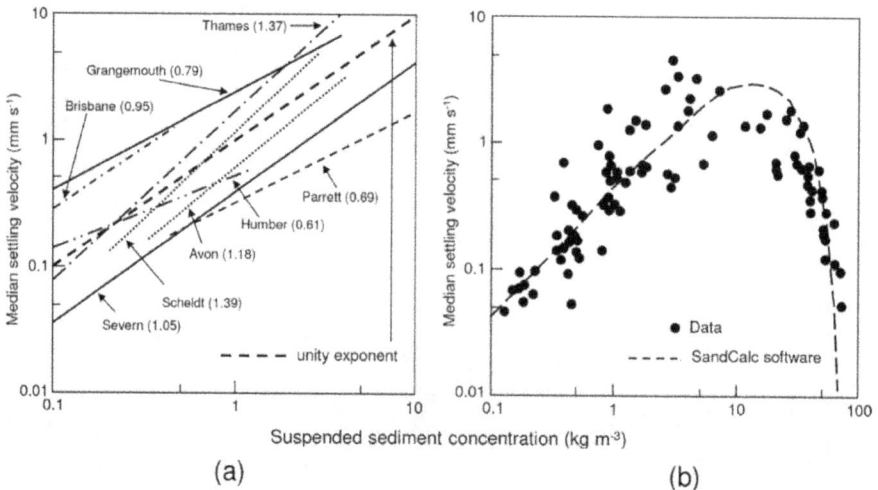

(a)   (b)

Figure 4.
Some examples of floc settling velocity measurements are shown. (a) Owen tube determinations of median settling velocity as a function of suspended sediment concentration for different estuaries; the bold dashed line represents an exponent of unity (reproduced with minor modifications from [49]). (b) Median settling velocity of Severn estuary mud as a function of SPM concentration; the Owen-tube data are taken from Odd and Roger [50], and the dashed line represents results based on the SandCalc software sediment-transport computational algorithm, which incorporates the hindered settling effect at high concentrations (reproduced with minor modifications from Soulsby [51]).

2.1.4.3 The van Leussen parameterisation

More recently, a number of authors have proposed simple theoretical formulae interrelating a number of floc characteristics that can then be calibrated using empirical studies. Such an approach has been used by van Leussen [19], who utilized a formula that modifies the floc settling velocity in still water by a floc growth factor, due to turbulence, and then reduces it by a turbulent floc disruption factor. The reference settling velocity (taken at low turbulent shear conditions), W_{s_0}, is then related to the SPM concentration (C) by a power law:

$$W_{s_0} = k \cdot C^m \qquad (1)$$

where k and m are empirical constants. The van Leussen is a qualitative simplification of a model originally developed for the sanitation industry [52], with only a limited number of interrelated parameters, and hence does not provide a complete description of floc characteristics within a particular sheared environment.

2.1.4.4 The Lick et al. parameterisation

A number of authors have attempted to observe how the floc diameter changes in turbulent environments. In particular, Lick et al. [53] derived an empirical relationship based on laboratory measurements made in a flocculator. They found that the floc diameter varied as a function of the product of the SPM concentration and the turbulence parameter as the turbulent shear rate, G:

$$D = c(C \cdot G)^{-d} \qquad (2)$$

where c and d are empirically determined values. However, this formulation provides no information on the important floc settling or floc dry mass properties.

2.1.4.5 The Manning and Dyer parameterization

The Manning Floc Settling Velocity (MFSV) algorithm for settling velocity [54] is based entirely on empirical observations made in situ using nonintrusive floc and turbulence data acquisition techniques in a wide range of estuarine conditions. The floc population size and settling velocity spectra were sampled using the video-based INSSEV instrument and LabSFLOC data.

The Manning-Dyer algorithms were generated by a parametric multiple regression statistical analysis of key parameters, which were generated from the raw, spectral floc data. Detailed derivations and preliminary testing of the floc-settling algorithms are described by Manning [26, 55]. Although the resulting empirical formulae are not presented in a fully dimensionless form, these formulae have the merit of being based on a large dataset of accurate, in situ settling velocity measurements (157 individually observed floc populations), acquired from different estuaries (Tamar, Gironde and Dollard) and different estuarine locations, such as the turbidity maximum and the intertidal zone.

The algorithms are based on the segregation of flocs into macroflocs ($D > 160$ μm [32]) and microflocs ($D < 160$ μm), which comprise the constituent particles of the macroflocs. This distinction permits the discrete computation of the mass settling flux (MSF) at any point in an estuarine water column.

Equations are given for: the settling velocity of the macrofloc fraction ($W_{s,macro}$), the settling velocity of microflocs (W_s) and the ratio of macrofloc mass to microfloc mass in each floc population—termed the SPM ratio [55]. This type of formulation gives a good compromise between the representation of physicochemical processes and computational simplicity. Eqs (3) and (4) describe $W_{s,macro}$ (mm s^{-1}) with inputs of SPM in mg l^{-1} and τ in Pa:

$$W_{s,macro} = 0.644 + 0.000471SPM + 9.36\tau - 13.1\tau^2 \tag{3}$$

for $0.04 < \tau < 0.6$ Pa

$$W_{s,macro} = 3.96 + 0.000346SPM - 4.38\tau + 1.33\tau^2 \tag{4}$$

for $0.6 < \tau < 1.5$ Pa. These equations require the input of a turbulent shear stress (τ) and an SPM concentration. These regression equations provide a realistic approximation to the field data. Graphical representations of the equations, together with the data, are presented in Manning and Dyer ([54]; see **Figure 5**). The Manning settling algorithm is valid for SPM concentrations in the range 10–8600 mg l^{-1} and shear stress values of $\tau < 2.13$ Pa, with extrapolation extending this range to 5–10 Pa.

An example of this is the implementation of the algorithm in a TELEMAC-3D numerical model of the Thames Estuary, UK [57], in which it was shown that the use of the Manning algorithm greatly improved the reproduction of observed distributions of SPM concentrations compared with the other formulations, both in the vertical and horizontal dimensions.

The Manning settling algorithms have been extended to cater for mixed sediment flocculation settling, including different ratios of mud:sand ([24, 47, 56]; see **Figure 5**). These algorithms are a major step forward in establishing a reliable estimate of the settling velocity. It has been developed based on a large and reliable dataset, it caters for the spectrum of hydrodynamic conditions that occur during a typical tidal cycle [58] (a feature often lacking in the settling terms of many estuarine sediment transport models) and has been shown to more accurately reproduce the distribution of suspended sediment compared with simpler settling models.

Soulsby et al. [59] has developed a more 'physics-based' version of the empirical model based on the Manning-Dyer formulation, called Soulsby-Manning 2013. It should be noted that for flocculation algorithms and models that include turbulence as a contributing variable, it is vital to ensure that the turbulence data are accurate, otherwise it has significant implications for the accuracy of the calculated floc settling characteristics.

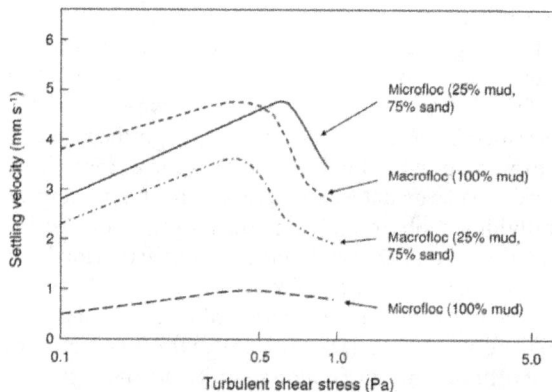

Figure 5.
Illustration of the settling velocities of macroflocs and microflocs, plotted against shear stress, for a mixed sediment suspension comprising a ratio of 25 per cent mud to 75 per cent sand and a pure mud suspension, all for a total SPM concentration of 5 gl^{-1} (modified from Figure 14 of Manning et al. [56]).

2.1.4.6 Complex population approaches

Lee et al. [60] applied a time-evolving two-class population balance equation (PBE) to determine the spatially and temporally changing distribution of fixed-size microflocs and size-varying macroflocs for bimodal floc distributions, with a fractal relationship between floc size and mass to derive the distribution of settling velocities. However, the authors felt that further intensive investigation of the aggregation and breakage kinetics would be required before their model was generally applicable when compared with the simpler approach of Manning and Dyer [54] and, presumably, Soulsby et al. [59].

Verney et al. [61] applied a time-evolving, multi-fraction model to determine the spatially and temporally changing distribution of the numbers of flocs in each size fraction, with a fractal relationship between floc size and mass to derive the distribution of settling velocities.

A relationship between the floc settling velocity and floc properties and fractal dimensions is given by Winterwerp [62]. A fractal approach has been used by Winterwerp [63] to solve a differential equation that simulates the time-varying representative floc diameter, from which floc density is derived from fractal considerations, and settling velocity obtained from a Stokes-like formula. Winterwerp et al. [64] also used a simplified fractal model to relate settling velocity to a turbulent shear parameter, the instantaneous concentration, and water depth.

The state-of-the-art model for floc structure is to assume a fractal structure. Many studies [65–67] indicate that the most sensitive parameters in a fractal model controlling the resulting settling velocity are the primary particle properties (primary particle diameter, density and their distributions) and the fractal dimension. For organic-rich particles, evidence suggests that the fractal dimension highly depends on stickiness [68].

Floc breakup in the existing size-class-based PBE formulation is modeled by assuming an invariant floc structure (e.g., fractal dimension ~ 2) and key properties, such as floc yield strength, are assumed constant over a wide range of floc size. The importance of modeling floc yield strength as a function of floc size has been demonstrated by Son and Hsu [67] via a fractal concept [69, 70]. Further details are provided in Section 3.1 below.

A fractal model is widely used to parameterize floc structure. Fractals are a mathematical simple approach, where scientists feel computationally 'comfortable' and therefore are happy to 'shoehorn' all flocs into this framework, and a fractal dimension of around 2 is often used. However, its applicability for heterogeneous sediments remains to be proved. Moreover, for flocs of high organic content, stickiness can be significantly enhanced due to the presence of extracellular polymeric substances (EPS) and transparent exopolymer particles (TEP). Field observations suggest that the fractal dimension for inorganic particles is larger than 2.0 while for organic rich flocs, it can be smaller than 2.0 [68]. There are also empirical formulations suggesting that the fractal dimension depends on floc size [65, 66].

Unlike Verney et al. [61], who use floc diameter, Maggi et al. [71] describe the floc population based on the number of primary particles in the flocs, which appears to make the incorporation of a variable fractal dimension straightforward. Moreover, Maggi et al. [71] adopt a sophisticated collisional efficiency closure that considers the effects of floc size and permeability.

2.1.5 Future floc modeling directions

Real flocs are multi component of different densities; even measuring real fractal dimensions is highly problematic. The emergence theory has been utilized in many disciplines (e.g. [72–74]) and provides a valid alternative and potentially more realistic approach for representing real multi-component mud floc structures. Both Cranford et al. [75], and Rietkerk and van de Koppel [76] have successfully adapted an emergence approach for application to natural biomaterials and ecosystems (respectively).

By utilizing an emergence framework for flocculation, at one end a simple fractal representation would still operate for basic, geometrically repeating simple floc structures (e.g. flocs composed from a single clay primary particle). Whilst as the flocs before more complex in structure, composition and geometry, and fractal theory become less representative, the emergence would adapt to a more suitable floc representation. It is envisaged that this new emergence approach [77] could cope better and more efficiently and realistically for real flocs at a wide range in resolution scales all using real image data at each scale. This will provide a level of error checks that are not supported by regular fractal approaches. Nonetheless, we are some way off from implementing this approach in a numerical floc model and more fundamental research on floc dynamics, properties and characteristics is required, in particular 2D and 3D floc imaging techniques (e.g. [27, 78–82]).

2.2 Grain-resolving simulations

Grain-resolving simulations are a powerful tool to obtain detailed, high-fidelity data of complex fluid-particle systems. Despite its rather large computational costs, recent advances in computational power have made it possible to perform grain-resolving simulations on scales that become relevant for sediment transport phenomena. The idea is to compute the trajectories of all the individual grains in a flow. Typically, the flow is computed on a Eulerian grid that is fixed in space and time. It can either be approximated by assuming a prescribed background flow or by solving the Navier–Stokes equations. The movement of the particles is then computed by their equations of motion in a Lagrangian sense, i.e. the particle is free to move within the entire computational domain. Hence, this representation is commonly referred to as the Euler–Lagrange approach.

Depending on their fluid-particle coupling, several schemes can be employed. If the particles are driven by a fluid flow but do not modify the flow field, the scheme is considered one-way coupled, whereas the fluid-particle mixture is two-way coupled if the flow is modified by the particle motion as well. In addition, momentum exchange of particles may be accounted for by means of collision and contact or any other particle related force. Using a two-way coupled simulation that accounts for particle-particle interactions is considered a fully-coupled scheme. In the context of the present study, attractive cohesive forces can contribute to the particle-particle interaction by binding grains into aggregates that are much larger than the individual primary particle. This process can be understood as flocculation. In the following, we will first review the governing equations to compute the particle motion in a Lagrangian sense and then proceed to present two examples to model flocculation of cohesive sediment to investigate aggregation and settling processes.

2.2.1 Computing the particle motion

Regardless of the background flow, we prescribe the motion of each primary cohesive particle i as a sphere moving with translational velocity $\mathbf{u}_{p,i}$ and angular velocity $\omega_{p,i}$. These are obtained from the Newton-Euler equations

$$m_{p,i}\frac{\mathrm{d}u_{p,i}}{\mathrm{d}t} = \underbrace{\oint_{\Gamma_{p,i}} \tau \cdot n \mathrm{d}A}_{F_{h,i}} + \underbrace{\frac{\pi D_{p,i}^3 \left(\rho_{p,i} - \rho_f\right)}{6} g}_{F_{g,i}} + \underbrace{\sum_{j=1, j\neq i}^{N_p} \left(F_{con,ij} + F_{lub,ij} + F_{coh,ij}\right)}_{F_{c,i}},$$

$$\tag{5}$$

$$I_{p,i}\frac{\mathrm{d}\omega_{p,i}}{\mathrm{d}t} = \underbrace{\oint_{\Gamma_{p,i}} r \times (\tau \cdot n)\mathrm{d}A}_{T_{h,i}} + \underbrace{\sum_{j=1, j\neq i}^{N_p} \left(T_{con,ij} + T_{lub,ij}\right)}_{T_{c,i}}, \tag{6}$$

where the primary particle i moves in response to the hydrodynamic force $F_{h,i}$, the gravitational force $F_{g,i}$ and the particle-particle interaction force $F_{c,i}$ which accounts for the direct contact force $F_{con,ij}$ in both the normal and tangential direction, as well as for short-range normal and tangential forces due to lubrication $F_{lub,ij}$ and cohesion $F_{coh,ij}$, where the subscript ij indicates the interaction between particles i and j. The hydrodynamic torque is denoted by $T_{h,i}$, while $T_{c,i}$ represents the torque due to particle-particle interactions, where we distinguish between direct contact torque $T_{con,ij}$ and lubrication torque $T_{lub,ij}$. Here, $m_{p,i}$ denotes the particle mass, $D_{p,i}$ the particle diameter, $\rho_{p,i}$ the particle density, ρ_f the fluid density, g the gravitational acceleration, $\Gamma_{p,i}$ the fluid–particle interface, τ the hydrodynamic stress tensor, N_p the total number of particles in the flow and $I_{p,i} = \pi\rho_{p,i}D_{p,i}^5/60$ the moment of inertia of the particle. Furthermore, the vector n represents the outward-pointing normal on the interface $\Gamma_{p,i}$, r is the position vector of the surface point with respect to the centre of mass of the particle.

Following Biegert et al. [83, 84], Zhao et al. [85, 86] represent the direct contact force $F_{con,ij}$ by means of spring-dashpot functions, while the lubrication force $F_{lub,ij}$ is accounted for based on lubrication theory [87] as implemented in Zhao et al. [85, 86]. The model for the cohesive force $F_{coh,ij}$ is based on the work of [88], which assumes a parabolic force profile, distributed over a thin shell of thickness h_{co} surrounding each particle:

$$\tilde{F}_{coh,ij} = \begin{cases} -4Co\dfrac{\zeta_{n,ij}^2 - h_{co}\zeta_{n,ij}}{h_{co}^2}n, & \zeta_{min} < \zeta_{n,ij} \leqslant h_{co}, \\[2ex] 0, & \text{otherwise}. \end{cases} \tag{7}$$

Here, ζ_n is the gap size in between particle surfaces, ζ_{min} is a limiter and Co is the cohesive number that has to be defined according to a given problem as will be detailed below. We remark that, based on Eqs. (5) and (6), the configuration of the primary particles within a floc can change with time in response to fluid forces, since the cohesive bonds are not rigid.

2.2.2 Aggregation

The flocculation process is strongly affected by the turbulent nature of the underlying fluid flow. Small-scale eddies modify the collision dynamics of the primary particles and hence the growth rate of the flocs, while turbulent stresses can result in the deformation and breakup of larger cohesive flocs. Hence, the dynamic equilibrium between floc aggregation and breakage is governed by a complex and delicate balance of hydrodynamic and inter-particle forces.

In the spirit of earlier investigations [89, 90], Zhao et al. [85] apply a simple model flow in order to investigate the effects of turbulence on the dynamics of cohesive particles. These authors consider the one-way coupled motion of small spherical particles in the two-dimensional, steady, spatially periodic cellular flow field commonly employed as initial condition for simulating Taylor-Green vortices (cf. **Figure 6a**), with fluid velocity field $u_f = (u_f, v_f)^T$

$$u_f = \frac{U_0}{\pi} \sin\left(\frac{\pi x}{L_0}\right) \cos\left(\frac{\pi y}{L_0}\right), \quad v_f = -\frac{U_0}{\pi} \cos\left(\frac{\pi x}{L_0}\right) \sin\left(\frac{\pi y}{L_0}\right), \tag{8}$$

where L_0 and U_0 represent the characteristic length and velocity scales of the vortex flow. By computing the fluid flow via this idealized flow field, the hydrodynamic torque $T_{h,i}$ can be omitted while the hydrodynamic force $F_{h,i}$ is generally replaced by a simple stokes drag force $F_{d,i} = -3\pi D_{p,i}\mu(u_{p,i} - u_{f,i})$ when simulations are one-way coupled in the sense that the particles do not modify the fluid flow. Here, $u_{f,i}$ indicates the fluid velocity evaluated at the particle center, μ the dynamic viscosity of the fluid.

The dynamics of the primary particles are characterized by the Stokes number $St = U_0 \rho_{p,i} D_{p,i}^2/(18L_0\mu)$, and the settling velocity $W_s = v_s/U_0$ where $v_s = (\rho_{p,i} - \rho_f)D_{p,i}^2 g/(18\mu)$ is the Stokes settling velocity of an individual primary particle i, as well as the cohesive number

$$Co = \frac{\max\left(\|F_{coh,ij}\|\right)}{U_0^2 L_0^2 \rho_f} = \frac{A_H(D_{p,i} + D_{p,j})}{32\lambda\zeta_0} \frac{1}{U_0^2 L_0^2 \rho_f}, \tag{9}$$

where the Hamaker constant A_H is a function of the particle and fluid properties, $\lambda = (D_{p,i} + D_{p,j})/40$ represents the range of the cohesive force and $\zeta_0 = (D_{p,i} + D_{p,j})/8000$ the characteristic distance. Representative values of A_H for common natural systems can be found in [88].

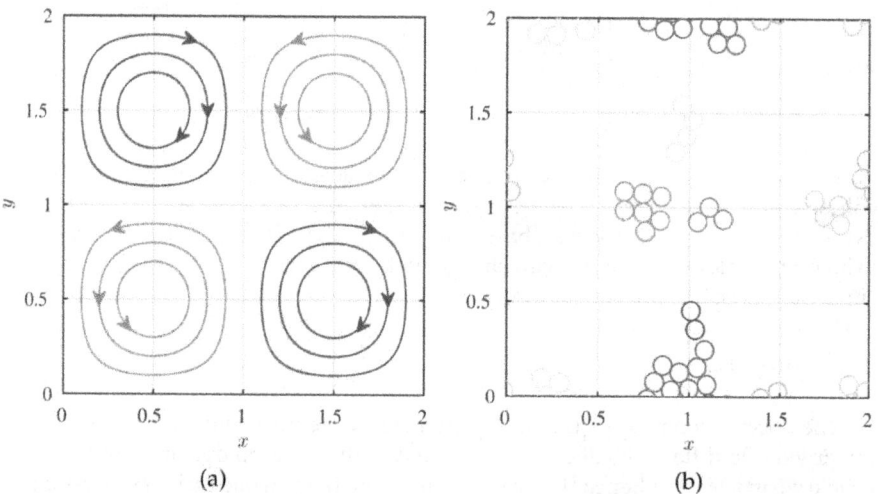

Figure 6.
(a) Streamlines of the doubly periodic background flow given by Eq. (8); (b) typical floc configuration with individual flocs distinguished by color (figure taken from [85].

Zhao et al. [85] employ a computational domain with periodic boundaries. All particles have identical diameters D_p and densities ρ_p. Initially they are at rest and separated, and randomly distributed throughout the domain. When the distance between two particles is less than $\lambda/2$, the particles are considered as part of the same floc and the number of flocs N_f is tracked as a function of time, with an individual particle representing the smallest possible floc (**Figure 6b**). Based on the simulation results, Zhao et al. [85] propose a new flocculation model to predict the temporal evolution of the floc size. For flocs of fractal dimension n_f, the mean floc size $\overline{D}_f$ is related to the average number of primary particles per floc $\overline{N}_{p,local} = N_p/N_f$,

$$\overline{D}_f = \left(\overline{N}_{p,local}\right)^{\frac{1}{n_f}} D_p \ , \tag{10}$$

$$\overline{N}_{p,local} = \frac{1}{\left(1/\overline{N}_{p,local,int} - 1/\overline{N}_{p,local,max}\right)e^{bt} + 1/\overline{N}_{p,local,max}} \ , \tag{11}$$

where $\overline{N}_{p,local,int}$ denotes the initial number of particles per floc and the average number of particles per floc during the equilibrium stage $\overline{N}_{p,local,max}$ is defined as,

$$\overline{N}_{p,local,max} = \begin{cases} N_p, & \overline{N}_{p,local,max} \geqslant N_p \ , \\ 8.5 a_1 St^{0.65} Co^{0.58} D_p^{-2.9} \phi^{0.39} \rho_s^{-0.49} (W_s+1)^{-0.38}, & \text{otherwise} \ , \end{cases} \tag{12}$$

where $\rho_s = \rho_p/\rho_f$ denotes the density ratio, ϕ the volume fraction of particles. The agglomeration rate $|b|$ with the constraint $b \leq 0$ is obtained by

$$b = \begin{cases} -0.7 a_2 St^{0.36} Co^{-0.017} D_p^{-0.36} \phi^{0.75} \rho_s^{-0.11} (W_s+1)^{-1.4}, & St \leqslant 0.7 \ , \\ -0.3 a_2 St^{-0.38} Co^{0.0022} D_p^{-0.61} \phi^{0.67} \rho_s^{0.033} (W_s+1)^{-0.46}, & St > 0.7 \ . \end{cases} \tag{13}$$

For the present cellular model flow the values $a_1 = a_2 = 1$ in Eqs. (12) and (13) yield optimal agreement with the simulation data with the fitting deviation of $\pm 30\%$. For real turbulent flows, a_1 and a_2 need to be determined by calibrating with experimental data. The new model, Eqs. (10)–(13), with a constant fractal dimension $n_f = 2$ for predicting the floc size has been employed and successfully validated with experimental data in our earlier work [85] (cf. **Figure 7**). The recent study by Zhao et al. [86] even goes beyond this assumption by showing that the fractal dimension becomes a function of the floc size when particles undergo flocculation in isotropic turbulence.

2.2.3 Hindered settling

It is well known that the settling behavior of a dense suspension differs substantially from the settling behavior of a single grain. Particles settling in a dense suspension induce a counterflow and experience friction by colliding with other particles. These processes yield the so-called hindered settling, which is substantially slower than the settling of an individual particle and depends on fluid and particle properties. Nevertheless, the Stokes settling velocity of an individual grain is still widely used to quantify the settling speed of sediment in particle-laden turbidity currents (e.g., [84]). Hence, constitutive equations to predict the settling speed as a function of the local flow conditions can enhance existing computational

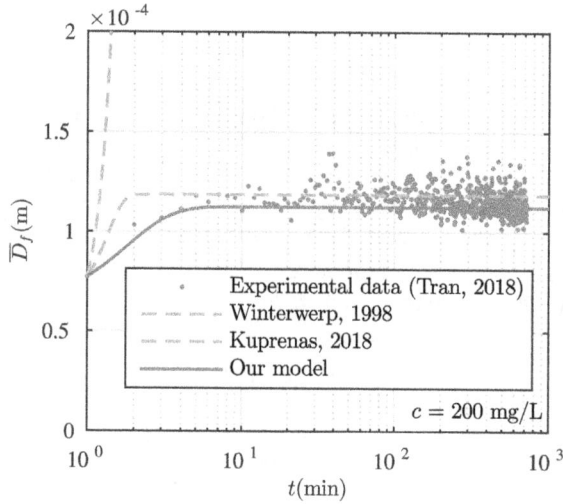

Figure 7.
Calibration of the empirical coefficients for the models of Winterwerp [62] ($k_A' = 1.35$ and $k_B' = 1.29 \times 10^{-5}$), Kuprenas et al. [91] ($k_A' = 0.45$ and $k_B' = 1.16 \times 10^{-6}$), and for our Eqs. (10)–(13) ($a_1 = 500$ and $a_2 = 35$); comparison between experimental data and predictions by the models. The experimental parameters are measured by Tran et al. [92], $D_p = 5$ µm, $\rho_p = 2650$ kg/m³, $\rho_f = 1000$ kg/m³, $\mu = 0.001$ Ns/m², the shear rate $G = 50$ s⁻¹, concentration $C = 200$ mg/L.

frameworks for the analysis of turbidity currents. To investigate the effects of the settling behavior of flocculating cohesive sediment by means of grain-resolving simulations [88, 93], it is important to not only account for frictional contact between particles in a dense suspension but also for the modifications of the fluid flow that is caused by settling particles displacing the fluid underneath them [94]. In this case, one needs to solve the Navier-Stokes equations for an incompressible Newtonian fluid:

$$\frac{\partial \mathbf{u}}{\partial t} + \nabla \cdot (\mathbf{uu}) = -\frac{1}{\rho_f} \nabla p + \nu_f \nabla^2 \mathbf{u} + \mathbf{f}_{IBM}, \tag{14}$$

along with the continuity equation

$$\nabla \cdot \mathbf{u} = 0, \tag{15}$$

where $\mathbf{u} = (u, v, w)^T$ designates the fluid velocity vector in Cartesian components, p denotes the pressure, ν_f is the kinematic viscosity, t the time, and $\mathbf{f}_{IBM}$ represents an artificial volume force introduced by the Immersed Boundary Method (IBM) [95, 96]. This volume force is introduced in the vicinity of the inter-phase boundaries to enforce a no-slip condition on the particle surface and to modify the fluid motion according to the particle motion. This measure also yields the hydro-dynamic forces and torques, $F_{h,i}$ and $T_{h,i}$ in Eqs. (5) and (6), respectively, as a direct result of this coupling scheme.

While this set of equations represents a fully coupled system, it is important to note that the relevant scales that define the system have changed compared to Section 2.2.2. In the scenario that investigates hindered settling of polydisperse cohesive sediment, the fluid flow is driven by moving particles. Hence, the relevant scale becomes $m_{50}g'$, where m_{50} is the mass of the median grain size and $g' = \left(\rho_p/\rho_f - 1\right)g$ is the specific gravity of the sediment. This scaling yields a modified cohesive number:

$$Co = \frac{\max\left(\|F_{coh,ij}\|\right)}{m_{50g'}} \tag{16}$$

and a characteristic particle Reynolds number $Re = D_{50}u_s/\nu_f$, where D_{50} is the median grain size and $u_s = \sqrt{g'D_{50}}$ is the buoyancy velocity of the sediment. These two non-dimensional numbers, Co and Re, fully define the physical system under consideration.

Particles that are placed in a tank will settle due to gravity thereby displacing the fluid and accumulate at the bottom of this tank. The induced counter flow as well as the particle-particle interaction yields frictional contacts and flocculation due to cohesive forces, which is the desired situation for hindered settling [88]. For small particle sizes, where cohesive forces remain relevant, Vowinckel et al. [88, 93] obtain a faster settling behavior for cohesive sediment as compared to its non-cohesive counterpart (**Figure 8**). During the settling process, the sediment will transform from a suspended state, where the weight is fully supported by the fluid pressure, to a deposited state, where the weight is supported through contact chains of the deposited sediment that extent all the way to the bottom of the tank [93]. This process is described by the effective stress concept, which states that the total stress, i.e. the submerged weight of sediment, is supported by either the particle pressure or the effective stress due to particle contact [97].

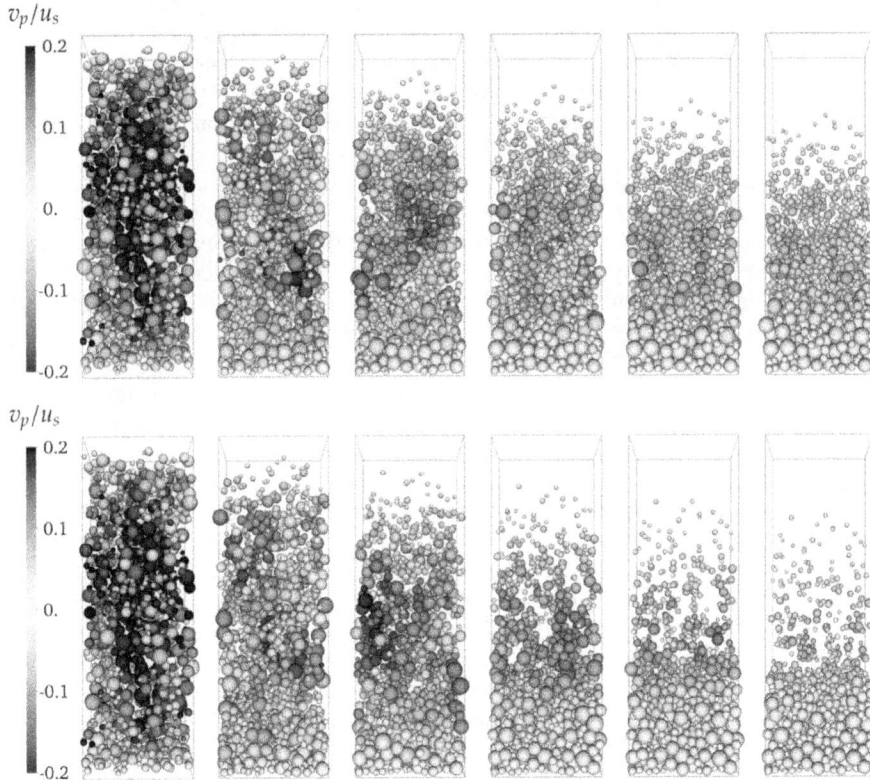

Figure 8.
Particle configurations during the settling and deposition process. Top row: Non-cohesive $(Co = 0)$ and bottom row: Cohesive $(Co = 5$ according to Eq. (16)). Left column: $t = 17.6\tau_s$, which corresponds to the time at which the particle phase has its maximum kinetic energy. Here, $\tau_s = D/u_s$ is the non-dimensional reference time. From left to right, the columns are separated by time intervals of $72.5\tau_s$. The gray shading reflects the vertical particle velocity (figure taken from [88]).

3. Coastal modeling

Coastal modeling typically refers to the modeling of regional scale (10–100 km^2) coastal and estuarine processes over a timescale of hours to months. Due to the large spatial and temporal scales that need to be covered, a Reynolds-averaged Navier-Stokes (RANS) model is adopted (e.g., [98–100]). Turbulence dissipation and mixing are parameterized with two-equation closure models via a diffusion process. Moreover, when surface waves are present, the individual wave-phase is often not resolved. The generation, transformation and dissipation of the random wave field are represented by a wave spectrum and solved by using the spectral wave action balance equation [101]. Wave-period-averaged wave statistics are then coupled with the coastal models. Consequently, the wave bottom boundary layer processes cannot be directly resolved and additional parameterizations are needed, such as the apparent roughness [102], i.e. the effect of the wave bottom boundary layer on the current resolved by the coastal model, wave-driven small-scale seabed morphological features (e.g., ripples), and the near-bed sediment transport processes (often called bedload or near-bed load). The suspended load transport for a range of non-cohesive sediment classes above the wave bottom boundary layer can be resolved in the coastal models via the conservation of mass. One of the main challenges for extending these coastal models for simulating cohesive sediment transport is the parameterization of settling velocity due to flocculation.

Since the recognition of flocculation in controlling the settling velocity of cohesive sediment in coastal and marine environments [12, 41], significant progress has been made, particularly in the physical parameters controlling the floc dynamics, floc size distribution and their relationships with settling velocity statistics. To name a few, the role of turbulent shear and particle concentration in determining the aggregation rate and the resulting floc size has been quantified [62, 103–105] (see also **Figure 1a**). Particularly, in many tidal boundary layers and laboratory experiments of homogeneous turbulence, the mean floc size is observed to be limited by the Kolmogorov length scale (e.g., [106, 107]). Moreover, the relationship between floc size population and settling velocity (or floc density) in both idealized and realistic conditions has been revealed (e.g., [108, 109]) and the fractal dimension has been applied to model these relationships [65, 66, 69, 71] (see also **Figure 3**). Finally, researchers have begun to understand complex floc characteristics in estuaries dominated by organic particles (e.g., [110]), high cohesion due to TEP (e.g., [111]) and the presence of sand (e.g., [112, 113], see also **Figure 5**). A more complete discussion on these observational-based empirical parameterizations is provided in Section 2.1.4.

Advancements have also been made in modeling flocculation processes of cohesive sediments. Following the summary presented in Section 2.1, this section focuses on a more in-depth discussion of the complex population approach. Pioneering work by Winterwerp [62, 94] established a robust single-size class (averaged floc size) flocculation modeling framework. This framework has been refined by Kuprenas et al. [91] to limit the floc size growth by the Kolmogorov length scale. A more sophisticated flocculation model [114] based on the population balance equations (PBE) has been incorporated into the Princeton Ocean Models (POM) by Xu et al. [115] to study the dynamics of Estuarine Turbidity Maximum (ETM). Recently, Sherwood et al. [116] incorporated the PBE flocculation model FLOCMOD by Verney et al. [2] into the Regional Ocean Modeling system (ROMS), which is part of the Coupled Ocean-Atmosphere-Wave-Sediment Transport Modeling System (COAWST, [98]). The model is used to study cohesive sediment transport in an idealized setting and a realistic application in the York River estuary. Through a direct simplification of the PBE type model, a tri-modal flocculation model was recently developed in the coastal model TELEMAC [117]. Last but not least, the empirical parameterization of the floc settling velocity MFSV algorithm (see Section 2.1.4) suggested by Manning and Dyer [54] has been incorporated into TELEMAC-3D [57], while more recently the nondimensional version of MFSV proposed by Soulsby et al. [59] has been incorporated into the Finite Volume Community Ocean Model (FVCOM, [99]). These closure models cover a wide range of flocculation physics incorporated or neglected, and the resulting computational cost also varies from minimal to very significant. It is worth to point out that besides the turbulence-averaged models discussed so far, a PBE formulation for flocculation dynamics has recently been incorporated into a turbulence-resolving large-eddy simulation model by Liu et al. [118] to study floc dynamics in the upper-ocean mixing layer subject to Langmuir turbulence.

3.1 Structure of continuum model

The Reynolds-averaged suspended sediment mass concentration $C(x_i, t)$ is solved by the conservation of mass in an advection-diffusion equation:

$$\frac{\partial C}{\partial t} + \frac{\partial u_i C}{\partial x_i} = \frac{\partial C W_s \delta_{i2}}{\partial x_i} + \frac{\partial}{\partial x_i}\left[(\nu + \nu_t)\frac{\partial C}{\partial x_i}\right] \tag{17}$$

where u_i is the fluid velocity, W_s is the settling velocity, ν is the fluid viscosity and ν_t is the turbulent (eddy) viscosity. The Kronecker delta δ_{ij} is used here with $j = 2$ representing the direction of gravitational acceleration. A challenge in modeling cohesive sediments is that the settling velocity, W_s, depends on the flocculation process. To include flocculation, there are generally two approaches [119]. The first approach is called the distribution based approach (e.g., [116, 120]), which directly models a bulk settling velocity as a variable due to flocculation. The simplest formulations are empirical formulas characterizing settling velocity as a function of mean floc size, sediment concentration, and turbulent shear rate [64]. Most notably, the framework provided by Winterwerp [62, 94] solves flocculation using a mean floc size, or number concentration of floc (having a mean floc size) that explicitly includes aggregation and breakup terms. Floc density and settling velocity can then be estimated by assuming a constant fractal dimension n_f. Such framework has been extended for a variable fractal dimension [67] and applied to model sediment resuspension in the Ems/Dollard estuary by Son and Hsu [121]. Flocculation process involved complex interaction between floc aggregation and breakup of different sizes, ranging from primary particles, microflocs, and macroflocs (see Section 2.1.2 for their definition). Using a mean value of floc property to describe these complex flocculation process may be too simplistic. For example, field observations suggest that natural flocs sometimes show a bimodal distribution (e.g., [18]). Maerz et al. [120] developed a flocculation model that solves the first moment of floc size distribution. Shen and Maa [119, 122] further use the quadrature of moments to model the evolution of floc size distribution. As discussed by Sherwood et al. [116], since settling velocity calculated from the distribution-based approach are based on the statistics of floc properties, the resulting bulk settling velocity must be a variable in time and space. Many existing coastal models utilize an efficient advection scheme that is designed for a constant and uniform settling velocity [115, 123]. Since these coastal models have been designed and routinely used to model suspended sediment (non-cohesive) transport of multiple size classes, it is more straightforward to extend the multiple-size class model into a PBE formulation to model flocculation processes.

3.2 Population balance equation

In the population balance formation, the sediment mass concentration C is partitioned into N classes and each class is represented by an index k, where k and N are positive integers. N must be a sufficiently large number to resolve the distribution. The sediment mass concentration in each class $c_k(x_i, t)$ is calculated by its mass conservation equation:

$$\frac{\partial c_k}{\partial t} + \frac{\partial u_i c_k}{\partial x_i} = \frac{\partial c_k w_{s,k} \delta_{i2}}{\partial x_i} + \frac{\partial}{\partial x_i}\left[(\nu + \nu_t)\frac{\partial c_k}{\partial x_i}\right] + G_{1,k} - L_{1,k} + \ldots, \qquad k = 1, 2, \ldots N$$

$$(18)$$

where $w_{s,k}$ is the settling velocity of class k. To include the flocculation processes, additional gain and loss terms, e.g., $G_{1,k}, L_{1,k}, \ldots$, due to aggregation of flocs (and primary particles) and breakup of flocs need to be modeled, which will be discussed later. It is important to point out that the flocculation processes only re-distribute the floc mass among different size classes and care must be taken to ensure the total sediment mass conservation when modeling these terms (i.e., these gain and loss terms must cancel each other after the summation of all size classes in Eq. (18)), including the numerical treatment (e.g., using a logarithmically distributed size class and a mass-weighted interpolation).

As mentioned before, the sediment mass concentration distribution can be described by floc size class [61] or number of primary particles in the floc [66]. Here, we focus on the more popular one using floc size class with each class having a floc diameter of $D_{f,k}$. The sediment mass concentration of each class can be related to number concentration of flocs in each size class $n_k(x_i, t)$ as:

$$c_k(x_i, t) = m_k n_k(x_i, t) \tag{19}$$

where m_k is the mass of a floc in size class k. Assuming a floc is formed by spherical primary particles of diameter D_p and density ρ^s, the mass of a primary particle can be calculated as

$$m_p = \frac{\pi}{6}\rho^s D_p^3. \tag{20}$$

By, furthermore, assuming that the floc structure follows a fractal relationship, we can calculate the mass of a floc as [69]:

$$m_k = m_p \left(\frac{D_{f,k}}{D_p}\right)^{n_f}. \tag{21}$$

Therefore, with the given fractal dimension and primary particle properties, m_k can be explicitly calculated for a given floc size class $D_{f,k}$. Following the fractal theory, the floc density of size class k can be readily calculated as [69].

$$\rho_k^f = \rho^w + (\rho^s - \rho^w)\left(\frac{D_{f,k}}{D_p}\right)^{n_f} \tag{22}$$

where ρ^w is the density of water (or seawater). With the known floc diameter and floc density, the settling velocity of the flocs in each size class can be calculated. The simple Stokes law is used here following Sherwood et al. [116]

$$w_{s,k} = \frac{\left(\rho_k^f - \rho^w\right)gD_{f,k}^2}{18\mu} \tag{23}$$

where μ is the dynamic viscosity of water (seawater).

It is worth noting that in the population balance formulation, the floc settling velocity of a particular size class is treated as a constant determined by the given fractal dimension, primary particle properties, and fluid properties. Therefore, it is more suitable for typical coastal modeling systems due to their numerical treatment of advection. In typical field conditions, a size-class based population balance formulation is reported to require at least 10–20 size classes [61, 116].

After proposing the appropriate gain and loss terms in Eq. (18), the full dynamics of floc transport, settling and re-distribution of sediment mass among all floc size classes due to flocculation can be modeled. In practice, some models solve a system of partial differential equation for the number concentration $n_k(x_i, t)$ directly by substituting Eq. (19) into Eq. (18) (e.g., Liu et al. [118]). For typical coastal models (e.g., Sherwood et al. [116]), the numerical treatment of flocculation and the advection-diffusion-settling processes are split into two steps. The zero-dimensional (homogenous turbulence condition without the advection, diffusion and settling terms) number concentration equations that only include the gain and loss terms are solved at every grid point over each (baroclinic) time step size of the coastal model to redistribute floc mass between different classes. Then, the newly calculated floc mass due to flocculation in each size class is updated by the advection-diffusion-settling equation (Eq. (18)) without the gain and loss (flocculation) terms. One advantage of such approach is that many zero-dimensional flocculation models developed elsewhere can be easily coupled into the coastal models. For example, Sherwood et al. [116] couple an existing zero-dimensional size-class based flocculation model FLOCMOD developed by Verney et al. [61] into the COAWST coastal modeling framework.

In this paper, the essential model formulations and closures of Verney et al. [61] are reviewed. The purpose here is not to discuss the model details, since they can be found in the cited references. Rather, this section is intended to bridge the discussions in Section 1 and Section 2 by focusing on the key model elements that are sensitive to the model results and hence may require more physical understanding. The governing equation for floc number concentration n_k (S.I. unit: m^{-3}) of size class k is written as

$$\frac{\partial n_k}{\partial t} = G_a(k) - L_a(k) + G_{bs}(k) - L_{bs}(k) \tag{24}$$

in which $G_a(k)$ and $L_a(k)$ represent the gain and loss of flocs in size class k due to aggregation, while $G_{bs}(k)$ and $L_{bs}(k)$ represent the corresponding gain and loss due to shear-driven breakup. Verney et al. [61] also include terms due to collision-induced breakup [70]. However, this mechanism is shown to be of minor importance for pure clay flocculation and it is not discussed here for brevity. The model assumes aggregation that is driven by turbulent shear and binary collision. Other known mechanisms driving aggregation, namely, the Brownian motion (important for particle smaller than 1 μm) and the differential settling are neglected by assuming that coastal and estuarine environments are dominanted by turbulent shear. The $G_a(k)$ and $L_a(k)$ terms are modeled as

$$G_a(k) = \frac{1}{2}\sum_{i+j=k} \alpha_{ij}A(i,j)n_i n_j \tag{25}$$

and

$$L_a(k) = \sum_i^N \alpha_{ik} A(i,k) n_i n_k \qquad (26)$$

where α_{ij} represents the collisional efficiency (dimensionless) and the shear-driven binary collision probability function is written as

$$A(i,j) = \frac{1}{6} G (D_i + D_j)^3 \qquad . \qquad (27)$$

The quantity $G_a(k)$ shown in Eq. (25) represents the gain of flocs in size k due to aggregation of smaller flocs at size class i and $j = k - i$ while $L_a(k)$ shown in Eq. (26) expresses the loss of flocs in size class k due to aggregation with flocs in other size classes. In Verney et al. [61], α_{ij} is set to be a constant $\alpha_{ij} = \alpha$ for simplicity. This point will be revisited later. Gain and loss of flocs in size class k due to breakup driven by turbulent shear are calculated as

$$G_{bs}(k) = \sum_{i=k+1}^N \Pi_{ki} B_i n_i \qquad (28)$$

and

$$L_{bs}(k) = B_k n_k. \qquad (29)$$

Essentially, gain of flocs $G_{bs}(k)$ in size class k due to fragmentation can only occur in floc size classes larger than k. The fragmentation probability function is written as

$$B_i = \beta_i G^{3/2} D_i \left(\frac{D_i - D_p}{D_p} \right)^{3-n_f} \qquad (30)$$

where β_i is the (dimensional: $m^{-1} s^{1/2}$) fragmentation rate and it is assumed to be a constant $\beta_i = \beta$ in Verney et al. [61]. More discussions on β_i will be given later. The breakup distribution function is defined as Π_{ki} and it characterizes different breakup scenarios. Our understanding on the shear-induced breakup scenarios is currently limited. Nevertheless, the breakup distribution functions Π_{ki} provided in Verney et al. [61] include binary (fragmentation of floc mass m_k into two small flocs of equal mass $m_k/2$), ternary (fragmentation of floc mass m_k into three smaller flocs, one floc of $m_k/2$ and two flocs of $m_k/4$), and erosion (fragmentation of floc mass m_k into $r + 1$ flocs, one larger floc and r smaller flocs of equal size).

As demonstrated in Verney et al. [61], their PBE-based flocculation model can predict several key features of floc dynamics observed in the field. For instance, the model is capable of reproducing the observed slower aggregation and more rapid fragmentation process (so-called clock-wise hysteresis of aggregation/fragmentation process) during a tidal cycle. As the floc size directly controls the settling velocity, capturing the hysteresis of floc aggregation/fragmentation is essential to further predict the net sediment fluxes during a tidal cycle. Moreover, Verney et al. [61] showed that a bimodal distribution of flocs often observed in the field can be reproduced by the PBE model by including a mix of different breakup distribution functions. Although the PBE-based flocculation models provide a promising modeling framework for cohesive sediment transport, there are limitations that require future investigations.

The most sensitive empirical parameters in FLOCMOD are the collisional efficiency α and the fragmentation rate β, which are often assumed to be constants. As demonstrated systematically by Verney et al. [61], to match the mean equilibrium floc size, there is no unique set of optimum α and β values and it is the ratio of α/β for these two empirical parameters that controls the mean equilibrium floc size. A similar finding is reported in the single size-class flocculation model of Winterwerp [62]. Hence, to fully benefit from the capability of PBE-based models that provide the temporal evolution of the full spectrum of floc sizes, the model calibration should go beyond just using equilibrium mean floc size. Sherwood et al. [116] provide a limited validation of Verney et al. [61] model for the temporal evolution of mean floc size. Such validation should be expanded for different data sets and for different floc size statistics in order to better constrain the empirical model parameters.

The sensitivity of the modeled mean floc diameter to the prescribed fractal dimension has also been discussed in Verney et al. [61]. More recently, Penaloza-Giraldo et al. [124] report that the temporal evolution of floc sizes (timescale to reach the equilibrium floc sizes) are sensitive to the fractal dimension n_f. As shown in Eq. (30), when n_f is smaller, the fragmentation probability function becomes larger, and hence it takes a shorter time to reach the equilibrium floc size. Even when α/β and n_f are calibrated to match the measured mean equilibrium floc size and flocculation timescale, it may not be straightforward to match the model results with measured data for the entire floc size distribution. Specifically, since the choice of breakup distribution function Π_{ki} is not constrained, the calibrated α/β and n_f may also depend on the Π_{ki}-function. In summary, the sensitivity study of Verney et al. [61] and preliminary findings reported by Penaloza-Giraldo et al. [124] imply that the following flocculation physics controlling the empirical parameters in the PBE warrant future studies:

3.2.1 Parameterizing the floc structure with a fractal dimension

To estimate floc mass and floc density in a given floc size, further assumptions on floc structures is needed. The state-of-the-art model for floc structure is to assume a fractal structure, which renders Eqs. (21) and (22) useful. In the PBE models for floc dynamics, fractal dimension directly affects the breakup term via the fragmentation probability function (see Eq. (30)). While most of the existing flocculation models assume a constant fractal dimension, examining the field and laboratory data by relating measured floc settling velocity and floc size (see Eqs. (22) and (23)) suggests that the fractal dimension may not be a constant, especially when considering the PBE equations are very sensitive to the prescribed fractal dimension value. Recent grain-resolving simulations (see Section 2.2.2 and the studies of [85, 86] referenced therein) also confirms that the fractal dimension depends on floc size. Researchers have proposed to model fractal dimension as a function of floc size [65, 71], however, whether it significantly improves the modeled flocculation processes remain to be proven.

3.2.2 More complete descriptions of collisional efficiency and fragmentation rate

As discussed by Hill and Nowell [125], the collision efficiency is practically treated as an empirical tuning parameter that parameterizes three main processes: encounter, contact and sticking. Encounter and contact are physical processes while sticking is associated with chemical-biological processes. From a physical perspective, only sticking efficiency is solely an empirical parameter. Maggi et al. [71] used a more complex collisional efficiency formulation that depends on the size and porosity of two colliding flocs. Through detailed laboratory experiments, Soos et al. [126] proposed a collisional efficiency formulation α_{ij} that depends on the size of the two colliding flocs and the turbulent shear rate. More systematic studies on the impact of collisional efficiency on the modeled flocculation in the PBE-based formulation are required.

While the existing studies mostly treat the fragmentation rate to be a model constant, physically the fragmentation rate β_i further depends on the floc breakage force F_y and the dynamic viscosity of the fluid [71]:

$$\beta_i = E\left(\frac{\mu}{F_y}\right)^{1/2} \tag{31}$$

with E an empirical constant. It is clear that the only way to justify that the fragmentation rate is a constant, i.e., $\beta_i = \beta$, is to assume that the floc breakage force F_y is independent of floc size, which theoretically is consistent with the fractal theory [69]. Indeed, in most cohesive sediment transport literature (e.g., [61, 71, 115]), F_y is assumed to be $F_y = \mathcal{O}(10^{-10})N$ following Winterwerp [62]. Since assuming floc structure follows the fractal theory (independent of the size of the aggregates) is an approximation, F_y may need to be considered as a variable in practice. Jarvis et al. [127] presented a review of many laboratory measurements of floc breakage force per unit area (it was called floc strength in their paper). Due to a wide range of cohesive sediment samples tested and different measurement techniques used, the quantitative results are not conclusive. However, it is clear that F_y/D_f^2 decreases as floc size increases. This conclusion does not exclude the possibility that F_y is a variable and it may be insensitive, or has a certain degree of dependence on floc size. More quantitative investigations on floc breakup force for different mineral composition and turbulence intensity are needed.

3.2.3 Improved physical understanding of breakup distribution functions

In the coastal sediment literature, detailed studies on floc breakup, particularly from the observational perspective, are rare. In the water quality literature, Jarvis et al. [127] provide some insights into the breakup distribution function. First, they discern the fragmentation mechanism, similar to the binary breakup, as the most likely scenario to occur when flocs are subjected to tensile stress acting across the floc. Secondly, the erosion mechanism in floc breakup is likely due to shear stress acting tangentially to the floc surface. Based on this argument, researchers hypothesize that the floc breakage types may depend on the ratio of floc size to the Kolmogorov length scale (smallest turbulent eddy size). However, the results are not conclusive and more comprehensive studies on how turbulent eddies interact with flocs and causing floc breakage are warranted.

4. Conclusions

We have presented an overview covering different types of floc analyses based on experimental measurements and grain-resolved simulations. These tools are currently emerging and show a very promising perspective to generate the data needed to account for unresolved cohesive sediment dynamics in continuum models with high fidelity. More work will be needed in the future to cover the

Physics of Cohesive Sediment Flocculation and Transport: State-of-the-Art Experimental...
DOI: http://dx.doi.org/10.5772/intechopen.104094

different aspects laid out in this chapter. Those are in particular, the effects of biofilms, the settling velocity of different types of flocs, as well as the aggregation and break-up efficiencies governing the exchange between different classes of PBE-type models. The knowledge to be gained can lead to a new generation of continuum models that enable simulations with predictive power for entire estuaries, which will bring inestimable advantages for these attractive settlement areas, both in economic terms and in terms of an increased quality of life.

Acknowledgements

BV gratefully acknowledges the support through the German Research Foundation (DFG) grant VO2413/2-1. KZ is supported by the National Natural Science Foundation of China through the Basic Science Center Program for Ordered Energy Conversion (51888103). LY, AJM, TJH and EM gratefully acknowledge supports by US National Science Foundation through a collaborative research between the University of Delaware (OCE-1924532) and the University of California Santa Barbara (OCE-1924655). AJM's contribution towards this research was partly supported by the US National Science Foundation under grants OCE-1736668 and OCE-1924532, TKI-MUSA project 11204950-000-ZKS-0002, and HR Wallingford company research FineScale project (ACK3013_62).

Conflict of interest

The authors declare no conflict of interest.

Nomenclature

$\alpha_{ij} = \alpha$	collisional efficiency
$\beta_i = \beta$	fragmentation rate
$\Gamma_{p,i}$	fluid-particle interface of particle i
$\delta_{i,j}$	Kronecker delta
ζ_0	characteristic distance
ζ_n	gap size in between particles
ζ_{min}	limiting gap size in between particles
λ	cohesive force range
μ	dynamic fluid viscosity
ν	kinematic fluid viscosity
ν_t	turbulent viscosity
Π_{ki}	breakup distribution function
ρ_f	fluid density
ρ_k^f	floc density of class k
$\rho_{p,i}$	density of particle i
ρ_s	density ratio
ρ^s	primary particle density
ρ^w	density of water
τ	turbulent shear stress
τ	hydrodynamic stress tensor
τ_s	nondimensional reference time
ϕ	particle volume fraction

$\omega_{p,i}$	Angular velocity vector of particle with index i
$A(i,j)$	Shear-driven binary collision probability function
A_h	Hamaker constant
a_1	model constant
a_2	model constant
b	agglomeration rate
B_i	fragmentation probability function
C	SPM concentration
c	empirical constant
c_k	sediment mass concentration of class k
Co	cohesive number
D	floc size
D_{50}	median grain size
$\overline{D}_f$	mean floc size
$D_{f,k}$	floc diameter of class k
D_p	primary particle diameter
$D_{p,i}$	diameter of particle i
E	empirical constant
$F_{c,i}$	particle-interaction force vector of particle with index i
$F_{coh,ij}$	cohesive force vector of particle i interacting with particle j
$F_{con,ij}$	normal contact force vector of particle i interacting with particle j
$F_{h,i}$	hydrodynamic force vector of particle with index i
$F_{g,i}$	gravitational force vector of particle with index i
f_{IBM}	artificial volume forced introduced by the IBM i
$F_{lub,ij}$	lubrication force vector of particle i interacting with particle j
F_y	floc breakage force
G	turbulence parameter / shear rate
$G_{1,k}$	mass gain due to aggregation of flocs
G_a	gain of flocs in a size class due to aggregation
L_{bs}	gain of flocs in a size class due to shear driven break-up
g	gravitational acceleration
g'	specific gravity
h_{co}	thickness of cohesive shell
$I_{p,i}$	moment of inertia of particle i
k	empirical constant
k	index of class for sediment mass concentration
k'_A	agglomeration constant
k'_B	breakup constant
L_0	characteristic length
$L_{1,k}$	mass loss due to breakup of flocs
L_a	loss of flocs in a size class due to aggregation
L_{bs}	loss of flocs in a size class due to shear driven break-up
m	empirical constant
m_k	mass of an individual floc in class k
N	number of classes for sediment mass concentration
n_f	fractal dimension
N_f	total number of flocs in the flow
n_k	number concentration of class k
N_p	total number of particles in the flow
$\overline{N}_{p,local}$	average number of primary particles per floc
m_{50}	mass of the median grain size

$m_{p,i}$	mass of particle i
m_k	mass of a floc
r	radial position vector
Re	Reynolds number
St	Stokes number
t	time
$T_{c,i}$	Torque vector of particle with index i due to particle-interaction
$T_{con,ij}$	collision torque vector of particle i interacting with particle j
$T_{h,i}$	hydrodynamic torque vector of particle with index i
$T_{lub,ij}$	lubrication torque vector of particle i interacting with particle j
U_0	characteristic velocity
$u_{f,i} = u$	fluid velocity vector
$u_{p,i}$	translational velocity vector of particle with index i
u_s	buoyancy velocity
v_s	Stokes settling velocity
W_s	settling velocity of microflocs
$w_{s,k}$	settling velocity of of class k
$W_{s,macro}$	settling velocity of macroflocs

Abbreviations

ADV	acoustic doppler velocimeter
COAWST	Coupled Ocean-Atmosphere-Wave-Sediment Transport Modeling System
DNS	direct numerical simulation
EPS	extracellular polymeric substance
ETM	estuarine turbidity maximum
FVCOM	finite volume community ocean model
IBM	immersed boundary method
INSSEV	IN-Situ SEttling Velocity instrument
LabSFLOC	laboratory spectral flocculation charactersitics
MFSV	manning floc settling velocity
MSF	mass settling flux
OBS	optical backscatter
PBE	population balance equation
PICS	particle imaging camera system
POM	Princeton ocean models
RANS	Reynolds-averaged Navier-Stokes
ROMS	regional ocean modeling systems
SB	size band
SPM	suspended particulate matter
TEP	transparent exopolymer particles

Author details

Bernhard Vowinckel[1,2*†], Kunpeng Zhao[2,3†], Leiping Ye[4†],
Andrew J. Manning[5,6,7,8,9,10,11*†], Tian-Jian Hsu[6†], Eckart Meiburg[2†]
and Bofeng Bai[3*†]

1 Leichtweiß-Institute of Hydraulic Engineering and Water Resources, Technische Universität Braunschweig, Braunschweig, Germany

2 Department of Mechanical Engineering, UC Santa Barbara, USA

3 State Key Laboratory of Multiphase Flow in Power Engineering, Xi'an Jiaotong University, Xi'an, China

4 School of Marine Sciences, Sun Yat-sen University, Zhuhai, China

5 HR Wallingford Ltd. Coasts and Oceans Group, Wallingford, UK

6 Center for Applied Coastal Research, Department of Civil and Environmental Engineering, University of Delaware, Newark, USA

7 Marine Physics Research Group, School of Marine Science and Engineering, University of Plymouth, Plymouth, UK

8 Stanford University, Stanford, California, USA

9 University of Florida, Gainesville, Florida, USA

10 Institute of Energy and Environment, University of Hull, Hull, UK

11 TU Delft, Delft, Netherlands

*Address all correspondence to: b.vowinckel@tu-braunschweig.de and andymanning@yahoo.com and bfbai@xjtu.edu.cn

† These authors contributed equally.

IntechOpen

References

[1] Nimmo Smith WAM, Katz J, Osborn TR. On the structure of turbulence in the bottom boundary layer of the coastal ocean. Journal of Physical Oceanography. 2005;**35**(1): 72-93

[2] Verney R, Lafite R, Brun Cottan JC. Flocculation potential of natural estuarine particles: The importance of environmental factors and of the spatial and seasonal variability of suspended particulate matter. Estuaries and Coasts. 2009;**32**:678-693

[3] Geyer WR, Lavery AC, Scully ME, Trowbridge JH. Mixing by shear instability at high Reynolds number. Geophysical Research Letters. 2010;**37**: L22607

[4] Foster DL, Beach RA, Holman RA. Turbulence observations of the nearshore wave bottom boundary layer. Journal of Geophysical Research, Oceans. 2006;**111**(C4):C04011

[5] Ozdemir CE, Hsu TJ, Balachandar S. Direct numerical simulations of transition and turbulence in Stokes boundary layer. Physics of Fluids. 2014;**26**:045108

[6] Thomson J, Schwendeman MS, Zippel SF, Moghimi S, Gemmrich J, Rogers WE. Wave-breaking turbulence in the ocean surface layer. Journal of Physical Oceanography. 2016;**46**(6): 1857-1870

[7] Lee JH, Monty JP, Elsnab J, Toffoli A, Babanin AV, Alberello A. Estimation of kinetic energy dissipation from breaking waves in the wave crest region. Journal of Physical Oceanography. 2017;**47**(5): 1145-1150

[8] Vowinckel B. Incorporating grainscale processes in macroscopic sediment transport models. Acta Mechanica. 2021;**232**:2023–2050. DOI: 10.1007/s00707-021-02951-4

[9] Kirby R. Suspended fine cohesive sediment in the Severn estuary and inner Bristol Channel, UK UKAEA Atomic Energy Research Establishment, Harwell. Energy 1986. p. 243

[10] van Leussen W. Fine sediment transport under tidal action. Geo-marine Letters. 1991;**11**(3):119-126

[11] Manning AJ, Bass SJ. Variability in cohesive sediment settling fluxes: Observations under different estuarine tidal conditions. Marine Geology. 2006; **235**(1–4):177-192

[12] Dyer KR. Sediment processes in estuaries: Future research requirements. Journal of Geophysical Research. 1989; **94**(C10):327-332

[13] Eisma D. Flocculation and de-flocculation of suspended matter in estuaries. Netherlands Journal of Sea Research. 1986;**20**(2–3):183-199

[14] Eisma D, Dyer K, Van Leussen W. The in-situ determination of the settling velocities of suspended fine-grained sediment—A review. In: Burt N, Parker R, Watts J, editors. Cohesive Sediments—Proceedings of INTERCOH Conference. Wallingford, England: John Wiley and Sons; 1997. pp. 17-44

[15] Law D, Bale A, Jones S. Adaptation of focused beam reflectance measurement to in-situ particle sizing in estuaries and coastal waters. Marine Geology. 1997;**140**(1–2):47-59

[16] Agrawal Y, Pottsmith H. Laser diffraction particle sizing in STRESS. Continental Shelf Research. 1994;**14** (10–11):1101-1121

[17] Graham GW, Nimmo Smith WAM. The application of holography to the analysis of size and settling velocity of suspended cohesive sediments.

Limnology and Oceanography: Methods. 2010;**8**(1):1-15

[18] Benson T, French J. InSiPID: A new low-cost instrument for in situ particle size measurements in estuarine and coastal waters. Journal of Sea Research. 2007;**58**(3):167-188

[19] van Leussen W. Estuarine Macroflocs and their Role in Fine-Grained Sediment Transport. The Netherlands: University of Utrecht; 1994

[20] Dearnaley M. Direct measurements of settling velocities in the Owen tube: A comparison with gravimetric analysis. Journal of Sea Research. 1996;**36**(1–2): 41-47

[21] Defossez J. Dynamique des macroflocs au cours de cycles tidaux, Mise au point d'un système d'observation. Rouen, France: VIL, Video In Lab. Mémoire de DEA, Université de Rouen; 1996

[22] Fennessy M, Dyer K, Huntley D. INSSEV: An instrument to measure the size and settling velocity of flocs in situ. Marine Geology. 1994;**117**(1–4):107-117

[23] Manning AJ, Dyer KR. The use of optics for the in situ determination of flocculated mud characteristics. Journal of Optics A: Pure and Applied Optics. 2002;**4**(4):S71

[24] Manning AJ, Baugh JV, Spearman JR, Pidduck EL, Whitehouse RJ. The settling dynamics of flocculating mud-sand mixtures: Part 1—Empirical algorithm development. Ocean Dynamics. 2011;**61**(2–3):311-350

[25] Manning A. LabSFLOC-2—The second generation of the laboratory system to determine spectral characteristics of flocculating cohesive and mixed sediments. In: HR Wallingford Technical Report (TR 156). Wallingford, UK: HR Wallingford; 2015

[26] Manning A. LabSFLOC—A laboratory system to determine the spectral characteristics of flocculating cohesive sediments. In: HR Wallingford Technical Report (TR 156). Wallingford, UK: HR Wallingford; 2006

[27] Ye L, Manning AJ, Hsu TJ, Morey S, Chassignet EP, Ippolito TA. Novel application of laboratory instrumentation characterizes mass settling dynamics of oil-mineral aggregates (OMAs) and oil-mineral-microbial interactions. Marine Technology Society Journal. 2018;**52**(6): 87-90

[28] Manning A, Schoellhamer DH. Factors controlling floc settling velocity along a longitudinal estuarine transect. Marine Geology. 2013;**345**:266-280

[29] Smith SJ, Friedrichs CT. Size and settling velocities of cohesive flocs and suspended sediment aggregates in a trailing suction hopper dredge plume. Continental Shelf Research. 2011; **31**(10):S50-S63

[30] ten Brinke WB. Settling velocities of mud aggregates in the Oosterschelde tidal basin (the Netherlands), determined by a submersible video system. Estuarine, Coastal and Shelf Science. 1994;**39**(6):549-564

[31] Fennessy M, Dyer K. Floc population characteristics measured with INSSEV during the Elbe estuary intercalibration experiment. Journal of Sea Research. 1996;**36**(1–2):55–62

[32] Manning AJ. Study of the effect of turbulence on the properties of flocculated mud. PhD thesis. UK: Institute of Marine Sciences, Faculty of Science, University of Plymouth; 2001

[33] Stokes GG. On the effect of the internal friction of fluids on the motion of pendulums. Transactions of the Cambridge Philosophical Society. 1851; **9**:8-106

[34] Manning AJ, Whitehouse RJS, Uncles RJ. Suspended particulate matter: the measurements of flocs. In: Uncles RJ, Mitchell S, editors. ECSA practical handbooks on survey and analysis methods: Estuarine and coastal hydrography and sedimentology, Chapter 8. Cambridge University Press; 2017. pp 211-260. DOI: 10.1017/9781139644426. ISBN 978-1-107-04098-4

[35] Dyer K, Cornelisse J, Dearnaley M, Fennessy M, Jones S, Kappenberg J, et al. A comparison of in situ techniques for estuarine floc settling velocity measurements. Journal of Sea Research. 1996;36(1–2):15-29

[36] Krone RB. A Study of Rheological Properties of Estuarial Sediments. Berkeley: Hydraulic Engineering Laboratory and Sanitary Engineering Research Laboratory, University of California; 1963

[37] Kolmogorov AN. The local structure of turbulence in incompressible viscous fluid for very large Reynolds numbers. Proceedings of the Royal Society of London Series A: Mathematical and Physical Sciences. 1890;1941(434):9–13

[38] Kolmogorov AN. Dissipation of energy in the locally isotropic turbulence. Doklady Akademii Nauk SSSR. 1941;32:16-18

[39] Mehta AJ, Lott JW. Sorting of fine sediment during deposition. In: Kraus NC, editor. Coastal Sediments '87, vol. I. New York: American Society of Civil Engineers; 1987. pp. 348–362

[40] Glasgow LA, Luecke RH. Mechanisms of deaggregation for clay-polymer flocs in turbulent systems. Industrial & Engineering Chemistry Fundamentals. 1980;19(2):148-156

[41] McCave IN. Size spectra and aggregation of suspended particles in the deep ocean. Deep Sea Research. 1984;31:329-352

[42] Alldredge AL, Gotschalk C. In situ settling behavior of marine snow 1. Limnology and Oceanography. 1988;33(3):339-351

[43] Manning A, Friend P, Prowse N, Amos C. Preliminary findings from a study of Medway Estuary (UK) natural mud floc properties using a laboratory mini-flume and the LabSFLOC system. Continental Shelf Research, BIOFLOW SI. 2007:1080-1095

[44] Overbeek JTG, Jonker G. Colloid Science Edited by HR Kruyt: Contributors-GH Jonker, J. Th. G. Overbeek. Amsterdam, The Netherlands: Elsevier; 1952

[45] Mikeš D, Manning AJ. Assessment of flocculation kinetics of cohesive sediments from the seine and Gironde estuaries, France, through laboratory and field studies. Journal of Waterway, Port, Coastal, and Ocean Engineering. 2010;136(6):306-318

[46] Pouët M.-F. La Clarification Coagulation—Flocculation Traitement de l'eau potable cours, EMA, option Environnement et Systèmes Industriels. 1997

[47] Manning AJ, Baugh JV, Spearman JR, Whitehouse RJ. Flocculation settling characteristics of mud: Sand mixtures. Ocean Dynamics. 2010;60(2):237-253

[48] Whitehouse RJS, Soulsby R, Roberts W, Mitchener HJ. Dynamics of Estuarine Muds. London: Thomas Telford Publications; 2000

[49] Delo EA, Ockenden MC. Estuarine muds manual. Report SR 309. Wallingford: HR Wallingford; 1992

[50] Odd NVM, Roger JG. An analysis of the behaviour of fluid mud in estuaries. HR Wallingford Technical Report, SR 84. 1986:94

[51] Soulsby RL. Methods for predicting suspensions of mud. In: HR Wallingford Technical Report (TR 104). Wallingford, UK: HR Wallingford; 2000

[52] Argaman Y, Kaufman WJ. Turbulence and flocculation. Journal of the Sanitary Engineering Division. 1970; **96**(2):223-241

[53] Lick W, Huang H, Jepsen R. Flocculation of fine-grained sediments due to differential settling. Journal of Geophysical Research: Oceans. 1993;**98** (C6):10279-10288

[54] Manning A, Dyer K. Mass settling flux of fine sediments in northern European estuaries: Measurements and predictions. Marine Geology. 2007;**245** (1–4):107-122

[55] Manning AJ. The observed effects of turbulence on estuarine flocculation. Sediment Transport in European Estuarine Environments: Proceedings of the STRAEE Workshop (WINTER 2004). Journal of Coastal Research. 2004; (Special issue 41):90-104

[56] Manning A, Spearman J, Whitehouse R, Pidduck E, Baugh J, Spencer K. Laboratory assessments of the flocculation dynamics of mixed mud: Sand suspensions. Chapter. 2013; **6**:119-164

[57] Baugh JV, Manning AJ. An assessment of a new settling velocity parameterisation for cohesive sediment transport modeling. Continental Shelf Research. 2007;**27**(13):1835-1855

[58] Mietta F, Chassagne C, Manning AJ, Winterwerp JC. Influence of shear rate, organic matter content, pH and salinity on mud flocculation. Ocean Dynamics. 2009;**59**(5):751-763

[59] Soulsby RL, Manning AJ, Spearman J, Whitehouse RJS. Settling velocity and mass settling flux of flocculated estuarine sediments. Marine Geology. 2013;**339**:1-12

[60] Lee BJ, Toorman E, Molz FJ, Wang J. A two-class population balance equation yielding bimodal flocculation of marine or estuarine sediments. Water Research. 2011;**45**(5):2131-2145

[61] Verney R, Lafite R, Brun-Cottan JC, Le Hir P. Behaviour of a floc population during a tidal cycle: Laboratory experiments and numerical modeling. Continental Shelf Research. 2011;**31**: S64-S83

[62] Winterwerp JC. A simple model for turbulence induced flocculation of cohesive sediment. Journal of Hydraulic Research. 1998;**36**(3):309-326

[63] Winterwerp H. On the dynamics of high-concentrated mud suspensions. Communications on Hydraulic and Geotechnical Engineering. 1999

[64] Winterwerp J, Manning A, Martens C, De Mulder T, Vanlede J. A heuristic formula for turbulence-induced flocculation of cohesive sediment. Estuarine, Coastal and Shelf Science. 2006;**68**(1–2):195-207

[65] Khelifa A, Hill PS. Models for effective density and settling velocity of flocs. Journal of Hydraulic Research. 2006;**44**:390-401

[66] Maggi F. Variable fractal dimension: A major control for floc structure and flocculation kinematics of suspended cohesive sediment. Journal of Geophysical Research. 2007;**112**:C07012

[67] Son M, Hsu TJ. The effect of variable yield strength and variable fractal dimension on flocculation of cohesive sediment. Water Research. 2009;**43**(14):3582-3592

[68] Engel A, Schartau M. Influence of transparent exopolymer particles (TEP) on sinking velocity of

Nitzschia closterium aggregates.
Marine Ecology Progress Series. 1999;**182**:
69-76

[69] Kranenburg C. The fractal structure
of cohesive sediment aggregates.
Estuarine, Coastal and Shelf Science.
1994;**39**(5):451-460

[70] McAnally WH, Mehta JA.
Significance of aggregation of fine
sediment particles in their deposition.
Estuarine, Coastal and Shelf Science.
2002;**54**:643-653

[71] Maggi F, Mietta F, Winterwerp JC.
Effect of variable fractal dimension on
the floc size distribution of suspended
cohesive sediment. Journal of
Hydrology. 2007;**343**:43-55

[72] Corning PA. The re-emergence of
"emergence": A venerable concept in
search of a theory. Complexity. 2002;
7(6):18-30

[73] Chialvo DR. Emergent complex
neural dynamics. Nature Physics. 2010;
6(10):744-750

[74] Halley JD, Winkler DA.
Classification of emergence and its
relation to self-organization.
Complexity. 2008;**13**(5):10-15

[75] Cranford SW, De Boer J, Van
Blitterswijk C, Buehler MJ. Materiomics:
An-omics approach to biomaterials
research. Advanced Materials. 2013;
25(6):802-824

[76] Rietkerk M, Van de Koppel J.
Regular pattern formation in real
ecosystems. Trends in Ecology &
Evolution. 2008;**23**(3):169-175

[77] Spencer KL, Wheatland JA,
Bushby AJ, Carr SJ, Droppo IG,
Manning AJ. A structure–function based
approach to floc hierarchy and evidence
for the non-fractal nature of natural
sediment flocs. Scientific Reports. 2021;
11(1):1-10

[78] Spencer KL, Manning AJ,
Droppo IG, Leppard GG, Benson T.
Dynamic interactions between cohesive
sediment tracers and natural mud.
Journal of Soils and Sediments. 2010;
10(7):1401-1414

[79] Wheatland JA, Bushby AJ,
Spencer KL. Quantifying the structure
and composition of flocculated
suspended particulate matter using
focused ion beam nanotomography.
Environmental Science & Technology.
2017;**51**(16):8917-8925

[80] Wheatland JA, Spencer KL,
Droppo IG, Carr SJ, Bushby AJ.
Development of novel 2D and 3D
correlative microscopy to characterise
the composition and multiscale
structure of suspended sediment
aggregates. Continental Shelf Research.
2020;**200**:104112

[81] Zhang N, Thompson CE,
Townend IH, Rankin KE, Paterson DM,
Manning AJ. Nondestructive 3D
imaging and quantification of hydrated
biofilm-sediment aggregates using X-
ray microcomputed tomography.
Environmental Science & Technology.
2018;**52**(22):13306-13313

[82] Ye L, Manning AJ, Hsu TJ. Oil-
mineral flocculation and settling
velocity in saline water. Water
Research. 2020;**173**:115569

[83] Biegert E, Vowinckel B, Meiburg E.
A collision model for grain-resolving
simulations of flows over dense, mobile,
polydisperse granular sediment beds.
Journal of Computational Physics. 2017;
340:105-127

[84] Biegert E, Vowinckel B, Ouillon R,
Meiburg E. High-resolution
simulations of turbidity currents.
Progress in Earth and Planetary
Science. 2017;**4**(1):33

[85] Zhao K, Vowinckel B, Hsu TJ,
Köllner T, Bai B, Meiburg E. An efficient

cellular flow model for cohesive particle flocculation in turbulence. Journal of Fluid Mechanics. 2020;**889**:R3

[86] Zhao K, Pomes F, Vowinckel B, Hsu TJ, Bai B, Meiburg E. Flocculation of suspended cohesive particles in homogeneous isotropic turbulence. Journal of Fluid Mechanics. 2021;**921**:A17

[87] Cox RG, Brenner H. The slow motion of a sphere through a viscous fluid towards a plane surface—II small gap widths, including inertial effects. Chemical Engineering Science. 1967; 22(12):1753-1777

[88] Vowinckel B, Withers J, Luzzatto-Fegiz P, Meiburg E. Settling of cohesive sediment: Particle-resolved simulations. Journal of Fluid Mechanics. 2019;**858**:5-44

[89] Maxey MR. The motion of small spherical particles in a cellular flow field. Physics of Fluids. 1987;**30**: 1915-1928

[90] Bergougnoux L, Bouchet G, Lopez D, Guazzelli E. The motion of solid spherical particles falling in a cellular flow field at low Stokes number. Physics of Fluids. 2014;**26**(9): 093302

[91] Kuprenas R, Tran D, Strom K. A shear-limited flocculation model for dynamically predicting average floc size. Journal of Geophysical Research-Oceans. 2018;**123**:6736-6752

[92] Tran D, Kuprenas R, Strom K. How do changes in suspended sediment concentration alone influence the size of mud flocs under steady turbulent shearing? Continental Shelf Research. 2018;**158**:1-14

[93] Vowinckel B, Biegert E, Luzzatto-Fegiz P, Meiburg E. Consolidation of freshly deposited cohesive and noncohesive sediment: Particle-resolved simulations. Physical Review Fluids. 2019;**4**(7):074305

[94] Winterwerp JC. On the flocculation and settling velocity of estuarine mud. Continental shelf research. 2002;**22**(9): 1339-1360

[95] Uhlmann M. An immersed boundary method with direct forcing for the simulation of particulate flows. Journal of Computational Physics. 2005; **209**(2):448-476

[96] Kempe T, Fröhlich J. An improved immersed boundary method with direct forcing for the simulation of particle laden flows. Journal of Computational Physics. 2012;**231**(9): 3663-3684

[97] Winterwerp JC, Kranenburg C. Fine Sediment Dynamics in the Marine Environment. Amsterdam, The Netherlands: Elsevier; 2002

[98] Warner JC, Armstrong B, He R, Zambon JB. Development of a coupled ocean-atmosphere-wave-sediment transport (COAWST) modeling system. Ocean Modelling. 2010;**35**:230-244

[99] Ge J, Shen F, Guo W, Chen C, Ding P. Estimation of critical shear stress for erosion in the Changjiang Estuary: A synergy research of observation, GOCI sensing and modeling. Journal of Geophysical Research, Oceans. 2015;**120**:8439-8465

[100] van Maren DS, Winterwerp JC, Vroom J. Fine sediment transport into the hyper-turbid lower Ems River: The role of channel deepening and sediment-induced drag reduction. Ocean Dynamics. 2015;**65**:589-605

[101] Booij N, Ris RC, Holthuijsen LH. A third generation wave model for coastal regions. Journal of Geophysical Research. 1999;**104**(C4):7649-7666

[102] Grant WD, Madsen OS. The continental-shelf bottom boundary layer. Annual Review of Fluid Mechanics. 1986;**18**(1):265-305

[103] Milligan T, Hill P. A laboratory assessment of the relative importance of turbulence, particle composition, and concentration in limiting maximal floc size and settling behaviour. Journal of Sea Research. 1998;**39**(3-4):227-241

[104] Keyvani SKA. Influence of cycles of high and low turbulent shear on the growth rate and equilibrium size of mud flocs. Marine Geology. 2014;**354**:1-14

[105] Tran C, Strom K. Suspended clays and silts: Are they independent or dependent fractions when it comes to settling in a turbulent suspension? Continental Shelf Research. 2017;**138**: 81-94

[106] Fugate DC, Friedrichs CT. Controls on suspended aggregate size in partially mixed estuaries. Estuarine, Coastal and Shelf Sciences. 2003;**58**:389-404

[107] Fettweis M, Francken F, Pison V, Van den Eynde D. Suspended particulate matter dynamics and aggregate sizes in a high turbidity area. Marine Geology. 2006;**235**:63-74

[108] Manning AJ, Dyer KR. A laboratory examination of floc characteristics with regard to turbulent shearing. Marine Geology. 1999;**160** (1-2):147-170

[109] Manning AJ, Bass SJ, Dyer KR. Floc properties in the turbidity maximum of a mesotidal estuary during neap and spring tidal conditions. Marine Geology. 2006;**235**(1-4):193-211

[110] Cartwright GM, Friedrichs CT, Sanford LP. In situ characterization of estuarine suspended sediment in the presence of muddy flocs and pellets. The Proceedings of the Coastal Sediments. 2011;**2011**:642-655

[111] Malpezzi MA, Sanford LP, Crump B. Abundance and distribution of transparent exopolymer particles in the estuarine turbidity maximum of

Chesapeake Bay. Marine Ecology Progress Series. 2013;**486**:23-35

[112] Manning AJ, Baugh JV, Spearman JR, Pidduck EL, Whitehouse RJS. Flocculation settlingCharacteristics of mud: Sand mixtures. Ocean Dynamics. 2010;**60**:237-253

[113] Cuthbertson AJS, Samsami F, Dong P. Model studies for flocculation of sand-clay mixtures. Coastal Engineering. 2018;**132**:13-32

[114] Xu F, Wang DP, Riemer N. Modeling flocculation processes of fine-grained particles using a size-resolved method: Comparison with published laboratory experiments. Continental Shelf Research. 2008;**28**:2668-2677

[115] Xu F, Wang DP, Riemer N. An idealized model study of flocculation on sediment trapping in an estuarine turbidity maximum. Continental Shelf Research. 2010;**30**:1314-1323

[116] Sherwood CR, Aretxabaleta AL, Harris CK, Rinehimer JP, Verney R, Ferré B. Cohesive and mixed sediment in the Regional Ocean Modeling system (ROMS v3.6) implemented in the coupled ocean–atmosphere–wave–sediment transport Modeling system (COAWST r1234). Geoscientific Model Development. 2018;**11**:1849-1871

[117] Shen X, Lee BJ, Fettweis M, Toorman EA. A tri-modal flocculation model coupled with TELEMAC for estuarine muds both in the laboratory and in the field. Water Research. 2018; **145**:473-486

[118] Liu J, Liang JH, Xu K, Chen Q, Chen Q, Ozdemir CE. Modeling sediment flocculation in Langmuir turbulence. Journal of Geophysical Research: Oceans. 2019;**124**:7883-7907

[119] Shen X, Maa JPY. Modeling floc size distribution of suspended cohesive sediments using quadrature method of

moments. Marine Geology. 2015;**359**: 106-119

[120] Maerz J, Verney R, Wirtz K, Feudel U. Modeling flocculation processes: Intercomparison of a size class-based model and a distribution-based model. Continental Shelf Research. 2011;**31**: S84-S93

[121] Son M, Hsu TJ. The effects of flocculation and bed erodibility on modeling cohesive sediment resuspension. Journal of Geophysical Research. 2011;**116**:C03021

[122] Shen X, Maa JPY. Numerical simulations of particle size distributions: Comparison with analytical solutions and kaolinite flocculation experiments. Marine Geology. 2016;**379**: 84-99

[123] Warner JC, Sherwood CR, Signell RP, Harris CK, Arango HG. Development of a three-dimensional, regional, coupled wave, current, and sediment-transport model. Computational Geosciences. 2008; **34**(10):1243-1260

[124] Penaloza-Giraldo J, Yue L, Ye L, Hsu TJ, Manning AJ, Meiburg E H, Vowinckel B. The Effect of Floc Strength in a Size Class-Based Flocculation Model. In AGU Fall Meeting Abstracts. San Francisco, USA; Dec 2020;**2020**:EP001-0020

[125] Hill PS, Nowell ARM. The potential role of large, fast-sinking particles in clearing nepheloid layers. Philosophical Transactions of the Royal Society of London A. 1990;**331**: 103-117

[126] Soos M, Wang L, Fox RO, Sefcik J, Morbidelli M. Population balance modeling of aggregation and breakage in turbulent Taylor–Couette flow. Journal of Colloid and Interface Science. 2007;**307**:433-446

[127] Jarvis P, Jefferson B, Gregory J, Parsons SA. A review of floc strength and breakage. Water Research. 2005;**39**: 3121-3137

Chapter 4

Rheology of Mud: An Overview for Ports and Waterways Applications

Ahmad Shakeel, Alex Kirichek and Claire Chassagne

Abstract

Mud, a cohesive material, consists of water, clay minerals, sand, silt and small quantities of organic matter (i.e., biopolymers). Amongst the different mud layers formed by human or natural activities, the fluid mud layer found on top of all the others is quite important from navigational point of view in ports and waterways. Rheological properties of fluid mud layers play an important role in navigation through fluid mud and in fluid mud transport. However, the rheological properties of mud are known to vary as a function of sampling location within a port, sampling depth and sampling location across the globe. Therefore, this variability in rheological fingerprint of mud requires a detailed and systematic analysis. This chapter presents two different sampling techniques and the measured rheological properties of mud, obtained from laboratory experiments. The six protocols used to measure the yield stresses are detailed and compared. Furthermore, the empirical or semi-empirical models that are commonly used to fit rheological experimental data of such systems are presented. The influence of different factors such as density and organic matter content on the rheological behavior of mud is discussed. The fluidic yield stress of mud samples was observed to vary from 0.2 Pa to 500 Pa as a function of density and organic matter content.

Keywords: Mud, rheology, density, yield stress, moduli, flow curve, protocol, organic matter, nautical bottom, cohesive sediment

1. Introduction

Mud beds, typically found at the bottom of rivers, lakes and in coastal areas, belong to the category of cohesive material. These deposits consist of water, clay minerals, sand, silt and organic matter (such as living microorganisms and in particular their excreted biopolymers) [1]. These mud beds are usually exposed to a continuous wave motion and disturbances produced by ship movement [1, 2], human actions such as dredging [3], natural climatic events and bioturbation [4]. The water column can be divided into different layers. In its large upper part, where mud particles are advected by currents and diffused by turbulent motion, mud is found as suspended particulate matter (SPM). Close to the bottom, different mud layers are found with increasing density as function of depth. These layers are defined as fluid mud (FM), pre-consolidated sediment (PS) and consolidated sediment (CS). Besides having different densities, these mud layers are known to have significantly different compositions and rheological fingerprints.

Fluid mud, the most important mud layer from a navigational perspective, is typically identified as a layer with a density of 1030–1300 kg m^{-3}, whereby hindered settling of particles plays a role due to the presence of flocs (i.e., combination of clay particles and organic matter) [5–7]. All mud layers, but particularly the fluid mud layer, display complex rheological behavior, i.e., combination of thixotropy, shear-thinning, two-step yielding behavior and viscoelasticity [8, 9]. The rheological/cohesive properties of mud are observed to vary as a function of solid fraction (or bulk density), type and concentration of organic matter, type of clay minerals and ionic concentration [10–17]. The thorough understanding of the rheological characteristics of mud, as a function of above-mentioned parameters, can help to estimate the strength, the flow and thickness of (fluid) mud in ports and water-ways. The quantification of the rheological properties for fluid mud also facilitates the definition of boundary conditions for sediment transport modeling, which in turn helps optimizing the dredging operations and defining the proper maintenance strategy for navigational channels [18–21]. However, in order to develop the appropriate in-situ techniques for measuring rheological properties, these characteristics need to be analyzed in laboratory beforehand. Therefore, in this chapter, following research questions are answered: How to efficiently collect the "undisturbed" mud sample? Which sediment properties are important for determining the rheological properties of mud? Which protocols are suitable for measuring rheological characteristics, i.e., yield stress of mud? Which empirical or semi-empirical model is appropriate to fit the rheological experimental data of mud, particularly for two-step yielding? How much comparable is the rheological signature of mud samples from different sources?

In this chapter, two different sampling techniques are presented to collect the "undisturbed" mud samples along with their important physical properties (Section 2). In Section 3, different protocols used to measure the rheological properties particularly yield stresses of mud are detailed and compared. Furthermore, the empirical or semi-empirical models that are commonly used to fit the rheological experimental data of mud are presented in Section 4. The influence of different factors such as density and organic matter content on the rheological behavior of mud is discussed in Section 5. In the end, the rheological properties, i.e., yield stress, of mud samples obtained from different ports are compared. In the present chapter, only laboratory experiments are presented.

2. Sampling techniques and physical properties of mud

In order to determine the physical and rheological characteristics of mud in the laboratory, appropriate sampling method needs to be applied. Two of the most commonly used sampling methods/equipment for mud are: (i) Van Veen grab sampler, and (ii) Frahmlot core sampler (see **Figure 1**). The criterion for selecting the suitable sampling method is based on the fact that the mud should be obtained in an "undisturbed" state with a naturally occurring density gradient profile, in order to estimate the properties of mud as close as possible to in-situ conditions. Core sampler is considered to meet this criterion well. Apart from collecting in-situ mud layers with different densities, another approach to study the effect of density on the rheological behavior is to dilute a consolidated mud layer, to obtain different samples with varying densities [9, 23]. However, the rheological characteristics of natural and diluted mud layers of same density are found to vary significantly from each other [24].

Figure 1.
(a) Van Veen grab sampler and (b) Frahmlot core sampler [22].

The bulk density (or water content) of the mud samples is usually estimated using the oven-drying method [25–28]. In short, the weight of the sample is recorded before and after heating at 105°C for 24 h. Using these weights and the density of water and minerals (i.e., 1000 and 2650 kg m^{-3}, respectively), the mean bulk density of the sample is obtained. The particle size distribution (PSD) of mud is typically investigated using (static) light scattering methods [17, 28, 29]. However, this technique also possesses some inherent drawbacks which are the facts that (i) the conversion between raw data and particle size is based on the assumption that the particles are spherical and have a homogeneous composition, (ii) the measurements are possible in a limited range of concentrations, and (iii) there can be a serious overestimation of the amount of large particles due to the mathematical smoothing of the PSD's by the manufacturer's software [30]. The total organic carbon (TOC) of mud samples is commonly analyzed by using loss-on-ignition method [31, 32], which is based on weighing the sample before and after heating at 430–500°C for 24 h. The total organic carbon is then estimated by loss in weight. All these sediment properties are known to significantly influence the rheological characteristics of mud samples.

3. Protocols for measuring rheological properties of mud

3.1 Yield stress

Mud can either behave as a solid-like material (i.e., elastic solid) at small stresses or as a liquid-like material above a critical value of stress, defined as yield stress. The nautical bottom for ports and waterways is typically defined on the basis of mud density [33], which does not account for the solid/liquid transition defined by the yield stress. Measurement of the yield stress of mud is, therefore, quite useful in defining the navigability of mud layers [15, 21]. The determination of yield stress is highly dependent on the selected rheological geometry and experimental method, as significantly different yield stress values can be obtained due to (i) the history of samples before analysis, (ii) differences in experimental methods, (iii) the use of different criteria for defining the yield point, (iv) different experimental timescales [34–39].

Several rheological geometries are available to perform the rheological analysis of mud including concentric cylinder (Couette), cone & plate (CP), parallel plate (PP) and vane. Cone & plate geometry has been observed to produce scattered rheological responses for mud samples due to the presence of large particles within the narrow gap between cone and plate and, hence, is not recommended for this kind of samples [40]. The remaining geometries can be used for analyzing mud samples but with certain benefits and limitations for each one. The differences between the geometries is illustrated in **Figure 2a** which shows the response of a FM mud layer in terms of elastic stress (= $G'\gamma$) as a function of applied oscillatory amplitude at 1 Hz for different geometries. It can be clearly seen that the mud sample exhibits a two-step yielding behavior (i.e., two distinct peaks in the response of elastic stress). The associated yield stresses are termed as "static" and "fluidic" yield stresses, and their values and dependence on amplitude are function of the used geometry. These two characteristic yield points can be associated to the breaking of floc network, re-organization and breakdown of flocs during shearing [40–43].

Figure 2b presents the yield stress values of different mud layers obtained by using the elastic stress method for different geometries. The results show that the highest yield stress values are obtained for parallel plate geometry, which may be attributed to the fact that this geometry induces the lowest disturbance in the sample while the plates are approached to confine the sample. However, this geometry is not very appropriate for analyzing the rheology of liquid-like samples, as the sample can flow out of the holder during shearing. Couette geometry is most suitable to analyze the rheological characteristics of mud ranging from very fluid to paste-like. For consolidated samples, however, it is preferable to use vane geometry, as the bob used in Couette geometry usually gets stuck during analysis of (very) dense mud samples.

In addition to different geometries, several rheological protocols have been reported in literature to determine the yield stress of mud. These protocols include shear rate ramp-up [29, 44], shear stress ramp-up [15, 17], and Claeys et al. protocol [45]. Shear rate/shear stress ramp-up methods are quite fast and easy to perform. On the other hand, Claeys et al. protocol is based on several cycles of selected shear rates (applying a shear, stop shearing, applying a shear ...) along with high shear rate steps in-between these cycles, with a total experimental time of about 15–20 min. **Figure 3** shows the pictorial representation of different experimental protocols that can be used to measure the yield stresses of mud.

The outcome of the different protocols in terms of shear stress as a function of shear rate or apparent viscosity as a function of shear stress for Port of Hamburg mud is shown in **Figure 4**. The values of yield stresses (static and fluidic) obtained

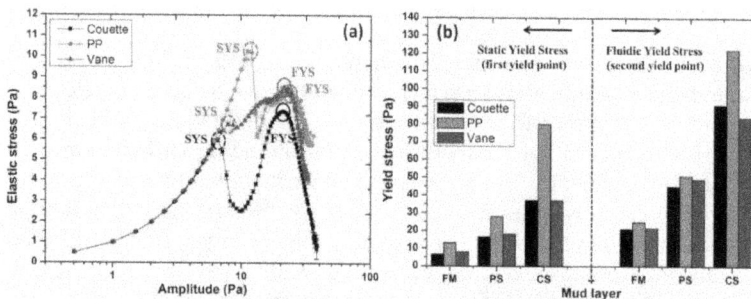

Figure 2.
(a) Elastic stress as a function of amplitude at 1 Hz using different geometries for FM layer obtained from Port of Hamburg. Solid line is just a guide for the eye. Bars represent standard deviation. Circles with dashed lines represent the static yield points (SYS) and circles with solid lines represent the fluidic yield points (FYS). (b) Static and fluidic yield stress values obtained from elastic stress method for different mud layers collected from Port of Hamburg using Couette, parallel plate and vane geometries. Reprinted from Ref. [40].

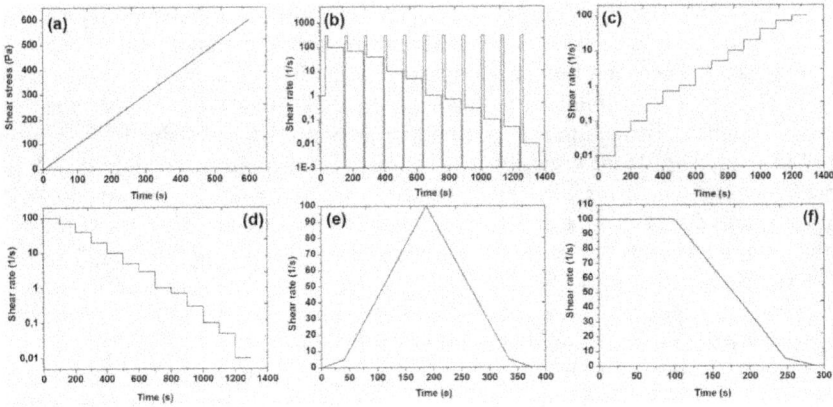

Figure 3.
Pictorial representation of the protocols (a) stress ramp-up, (b) Claeys et al. protocol, (c) increasing equilibrium flow curve (EFC), (d) decreasing equilibrium flow curve (EFC), (e) shear rate ramp up and ramp down (CSRT) and (f) pre-shear test. Reprinted from Ref. [46].

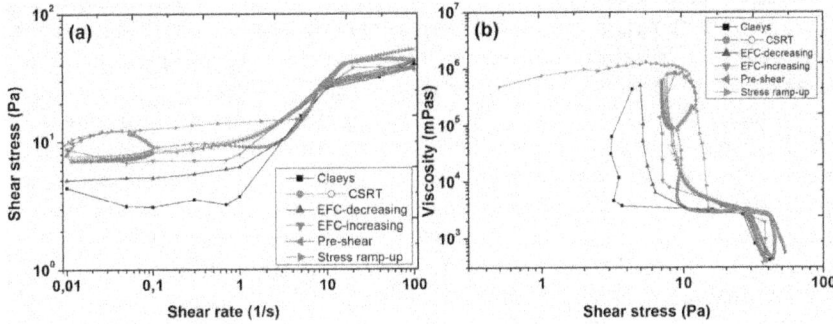

Figure 4.
(a) Shear stress as a function of shear rate and (b) apparent viscosity as a function of shear stress for mud sample collected from port of Hamburg using Couette geometry; solid symbols in CSRT protocol represent the ramp-up and the empty symbols represent the ramp-down; solid lines are just the guide for the eye. Reprinted from Ref. [46].

Method	Static Yield Stress (Pa)	Fluidic Yield Stress (Pa)
Claeys protocol	3.1–4.4	26
CSRT-ramp up	9.0–12.3	40
CSRT-ramp down	7.6	29
EFC-decreasing	5.2	26
EFC-increasing	7.1	38
Pre-shear	7.1	27
Stress ramp-up	11	40

Table 1.
Static and fluidic yield stress values of mud sample from port of Hamburg obtained from viscosity declines with Couette geometry for different protocols.

from the viscosity declines of these curves (see **Figure 4b**) are presented in **Table 1**. It can be seen from **Table 1** that higher yield stress values are obtained from stress ramp-up test, ramp-up step of shear rate ramp up and ramp down (CSRT) test and increasing equilibrium flow curve (EFC) test. This is linked to the fact that these methods deform mud samples from an almost undisturbed state to an almost fully disturbed state. These methods are, therefore, suitable to measure the yield stresses of mud close to in-situ conditions. However, the determination of a static yield point is somehow difficult in case of ramp-up step of CSRT test (due to the scattering of the initial apparent viscosity as function of stress points, see black curve in **Figure 4b**) and increasing EFC test is a somehow lengthy test ($\sim$ 20 min, see **Figure 3c** and **d**). Therefore, it is recommended to use stress ramp-up test for analyzing the yield stresses of mud for ports and waterways applications. The yield stress values obtained from the rest of the methods (i.e., Claeys et al., decreasing EFC, pre-shear and ramp-down step of CSRT) are quite lower, which indicates the extensive structural breakdown of the samples during analysis.

In literature, several terminologies have been used to represent the two yield points for mud. The correspondence between these terminologies is presented in **Table 2**.

3.2 Thixotropy and structural recovery

Thixotropy, a commonly observed rheological fingerprint of suspensions, is defined as a phenomenon in which the viscosity of the system is both shear rate and time dependent. Therefore, a thixotropic material shows a time dependent viscosity (decreasing as a function of time; when the viscosity is increasing as a function of time the material is said to be rheopectic) after applying/stopping shear rate [49]. Typically, the thixotropic behavior of mud is determined by performing a shear rate ramp-up followed by a constant high shear step and then a shear rate ramp-down step [17, 44]. The area of the hysteresis loop formed between the upward and downward curve quantifies the thixotropic behavior of the material [50]. Multiple thixotropic loops can also be produced for the same sample without allowing for any delay between each loop, in order to understand the thixotropic behavior of mud after extensive structural breakdown [17].

In addition to thixotropy, the structural recovery of mud after extensive shearing is also interesting to analyze by using small amplitude oscillatory rheological measurements. A three step experimental protocol has been reported in literature to quantify the structural recovery of mud after steady pre-shearing [51]. In short, the first step of the protocol involves the application of an oscillatory amplitude within the linear viscoelastic (LVE) regime and recording the moduli as a function of time (i.e., resting step). This step provides the initial moduli values before the pre-shearing step and also eliminates the disturbances created by the geometry. In a second step, a constant high shear rate is applied for a time interval which is enough to completely destroy the structure of the sample (i.e., pre-shear step). The last step provides the key information about the structural recovery of mud after

Shakeel et al. [17]	Toorman [47]	Toorman [48]	Wurpts & Torn [15]	Claeys et al. [45]
Fluidic yield stress		Static yield stress	Criterion for navigation	Undrained shear strength
ΔYS = (Fluidic – Static)	Bingham yield stress	Dynamic yield stress		Bingham yield stress
Static yield stress	(True) yield stress		Yield stress	

Table 2.
Correspondence between different yield stress terminologies reported in literature for mud.

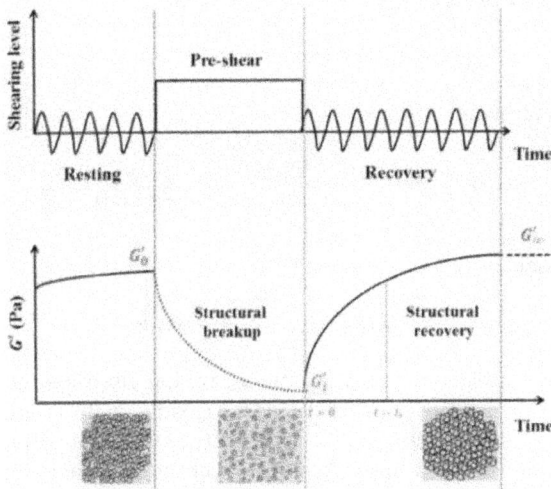

Figure 5.
Schematics of the experimental protocol employed for the structural breakup and recovery in mud samples.
Reprinted from Ref. [51].

pre-shearing, again by applying the oscillatory amplitude within LVE regime and recording the moduli as a function of time (i.e., structural recovery step). The schematic representation of the experimental protocol is shown in **Figure 5**.

3.3 Moduli

Apart from above mentioned rheological properties of mud, the estimation of moduli (storage and loss) within LVE regime (i.e., without significantly affecting the structure of the material) as a function of frequency provides useful information about the strength of the material. Preliminary oscillatory amplitude sweep experiments are usually performed to determine the LVE regime (where the response of material is independent of applied stress/strain). The frequency sweep tests are then performed by selecting the appropriate amplitude of oscillation within LVE regime and the desired frequency range [17, 29]. The results can either be plotted as moduli (storage and loss) vs. frequency or complex modulus and phase angle as a function of frequency. In addition to small amplitude oscillatory experiments, large amplitude oscillatory tests have also been reported in literature to analyze the nonlinear response of mud in terms of stress waveform and Lissajous pattern [41].

4. Rheological modeling of mud

4.1 Flow curve

In literature, the rheological behavior (i.e., flow curve) of mud has been fitted with numerous empirical or semi-empirical models including Bingham [52], Herschel-Bulkley [53], Worrall-Tuliani [54], and Toorman model [48], given as follows:

Bingham model:

$$\tau = \tau_B + K\dot{\gamma} \tag{1}$$

Herschel-Bulkley model:

$$\tau = \tau_0 + K\dot{\gamma}^n \tag{2}$$

Worrall-Tuliani model:

$$\tau = \tau_0 + \mu_\infty\dot{\gamma} + \frac{(\mu_0 - \mu_\infty)\dot{\gamma}}{1 + \frac{\mu_0 - \mu_\infty}{\tau_\infty - \tau_0}\dot{\gamma}} = \tau_0 + \mu_\infty\dot{\gamma} + \frac{\Delta\mu\dot{\gamma}}{1 + \beta\dot{\gamma}} \tag{3}$$

Toorman model:

$$\tau = \lambda\tau_0 + (\mu_\infty + c\lambda + \beta\tau_0\lambda_e)\dot{\gamma} \tag{4}$$

where K is the consistency index, μ_0 and μ_∞ are the viscosities at lower and higher shear rates, n is the flow behavior index, τ_0 is the yield stress and τ_∞ and τ_B are the Bingham yield stresses at high shear, λ is the structural parameter which varies between 0 and 1, λ_e is the equilibrium structural parameter and β is the ratio of breakdown and aggregation parameter. However, all these models are suited to fit the experimental data of flow curve with only single step yielding.

4.1.1 Empirical model for two-step yielding

It has been reported in literature that mud samples usually exhibit a two-step yielding phenomenon [17, 41, 51], which is associated with two yield stresses. These two yield points depict the transition between a fully structured sample (i.e., interconnected network of flocs), partially structured sample (i.e., mobile flocs) and almost fully disturbed sample (i.e., smaller flocs or particles). Therefore, the shear stress as a function of shear rate for the whole investigated range can be written as a sum of two functions, which represent the two yield regions, given as:

$$\tau = \alpha\tau_{stat} + (1 - \alpha)\tau_{fluid} \tag{5}$$

where α is a step function given by:

$$\alpha = 1 - \frac{1}{1 + \exp\left(-k(\dot{\gamma} - \dot{\gamma}_0)\right)} \tag{6}$$

where $\dot{\gamma}_0$ represents the shear rate at which the transition between the two regions occurs and its sharpness varies as a function of parameter k. The stress function for the first yield region can be written as follows:

$$\tau_{stat} = \frac{\tau_s}{1 + \dot{\gamma}_s/\dot{\gamma}} \tag{7}$$

The shear stress τ_s represents the first yield point typically known as static yield point and the shear rate $\dot{\gamma}_s$ can be used to control the curvature of this stress function. The stress function for the second yield region can be given by:

$$\tau_{fluid} = \tau_s + \frac{\tau_f}{1 + \left(\left(\dot{\gamma}_f - \dot{\gamma}_0\right)/(\dot{\gamma} - \dot{\gamma}_0)\right)^d} + \mu_\infty(\dot{\gamma} - \dot{\gamma}_0) \tag{8}$$

The shear stress τ_f represents the second yield point typically known as fluidic yield point and the parameter d can be used to tune the sharpness of this function.

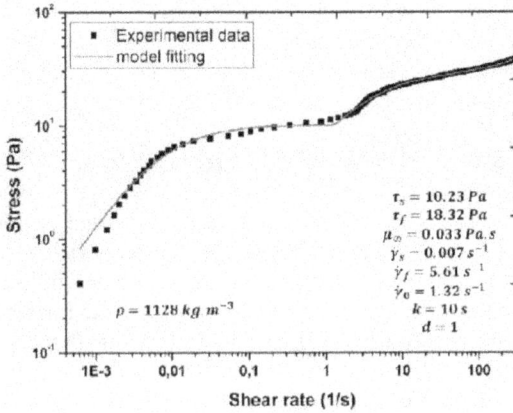

Figure 6.
Shear stress as a function of shear rate for mud sample from port of Hamburg obtained by performing stress ramp-up test using Couette geometry. The solid line represents the two-step yielding model fitting Eq. (5). Reprinted from Ref. [46].

Figure 6 shows the fitting of experimental data obtained for the mud sample from Port of Hamburg using Eq. (5) along with the values of fitting parameters.

4.2 Structural recovery

The experimental data of structural recovery tests (i.e., storage modulus as a function of time) can be easily fitted with the stretched exponential function [51] given as follows:

$$\frac{G'}{G'_0} = \frac{G'_i}{G'_0} + \left(\left(\frac{G'_\infty - G'_i}{G'_0} \right) \left(1 - \exp\left[-\left(\frac{t}{t_r} \right)^d \right] \right) \right) \qquad (9)$$

where G' represents the storage modulus recorded as a function of time after pre-shear step, G'_0 denotes the initial storage modulus value before pre-shear step, G'_i is the storage modulus value recorded immediately after pre-shear step, and d is the stretching exponent. The two remaining parameters, i.e., G'_∞ and t_r represent the equilibrium storage modulus of the system and the characteristic time of recovery (the time required to attain 63% of equilibrium storage modulus), respectively. These two parameters provide the strength of recovered structure and the time required to regain its strength, which is useful information for ports and waterways applications, where the mud is disturbed by natural or human activities in the ports.

5. Factors affecting rheological properties of mud

5.1 Density

Density (i.e., solid content and water content) is an important characteristic of mud which can significantly affect their rheological behavior, such as yield stress, thixotropy and moduli. In natural environment, the density gradient in mud is usually created by wave motion or human activities along with settling of particles. In literature, several researchers have reported the rheological characteristics of mud as a function of solid content or density [9, 17, 23, 26, 27, 29, 55]. For instance, the mud obtained from Lianyungang, China showed an exponential increase in yield stress as a function of volume concentration of particles [29]. Likewise, the similar exponential relation between yield stress and solid content or water content of mud has also been reported by other researchers for samples obtained from different ports [17, 26, 27, 55].

However, this correlation between yield stress and density is highly dependent on the mud composition. For example, the (fluidic) yield stress values as a function of density for mud samples collected from different locations of port of Hamburg is shown in **Figure 7a**. It can be seen that the dependence of yield stress on mud density is significantly different for the samples collected at different locations. In order to further quantify this difference, both the fitting parameter 'a' for the power law relation given in the caption of **Figure 7a** and the total organic carbon (TOC) are plotted as function of different locations (**Figure 7b**). There is clearly a correlation between the TOC content and the fitting parameter 'a'. This behavior suggests that the yield stress of mud is strongly dependent both on TOC and mud density, as already reported in literature [15, 16].

Several researchers have reported the rheological characteristics of mud as a function of density either by collecting natural mud layers with varying density [17] or by diluting dense mud samples [9, 23]. However, it has been observed that the natural and diluted mud layers display significantly different rheological properties [24] (see **Figure 8**), which may again be linked to the composition of each mud layer, procedure of dilution, etc.

Apart from yield stress, other rheological properties including moduli, thixotropy, structural recovery, etc. are also strongly dependent on the density of mud samples [24, 51]. For instance, the structural recovery, observed by using above mentioned protocol (Section 3.2), for mud samples collected from different locations and different depths is shown in **Figure 9a** and **b**, respectively. From the figure, it is found that the structural recovery (i.e., moduli values) of mud is highly dependent on the mud layer, and position in the harbor [51]. Hence, density of mud is a critical parameter particularly for describing their rheological characteristics, however, for defining nautical bottom in ports only density is not enough and other parameters also need attention from the researchers.

5.2 Organic matter content (TOC) and its degradation

The presence of organic matter in mud usually hinders the settling of particles and can help to form fluid mud layers, in addition to the natural wave motion or

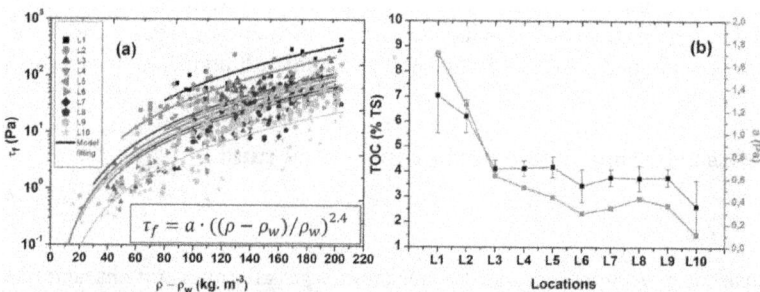

Figure 7.
(a) Fluidic yield stress (τ_f) as a function of excess density ($\rho - \rho_w$) for mud samples from different locations of port of Hamburg. The solid lines represent the power law fitting with one fitting parameter 'a'. ρ_w represents the density of water. L1 to L10 represent the locations from river towards sea side in the Port of Hamburg. (b) Fitting parameter 'a' and TOC as a function of different locations of port of Hamburg. Reprinted from Ref. [46].

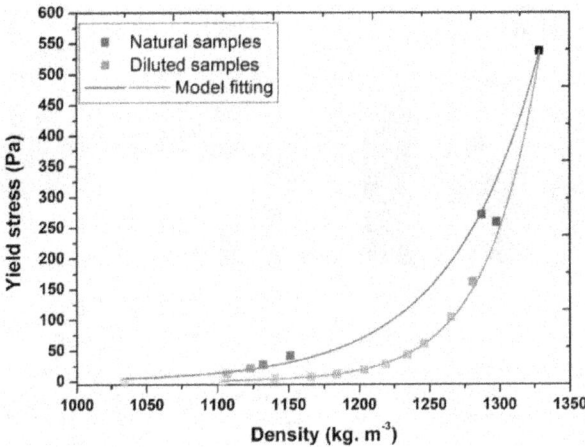

Figure 8.
Fluidic yield stress values vs. bulk density for natural and diluted mud layers from Port of Hamburg. Solid lines represent the power law fitting. Reprinted from Ref. [24].

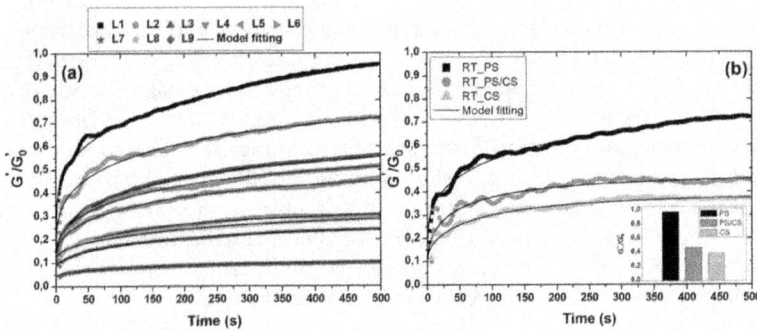

Figure 9.
(a) Normalized storage modulus (G'/G_o') as a function of time for pre-consolidated (PS) sediment samples obtained from different locations of port of Hamburg and pre-sheared at $100\ s^{-1}$ using Couette geometry, L1 to L10 represent the locations from river towards sea side in the port of Hamburg with similar densities. (b) Normalized storage modulus (G'/G_o') as a function of time for different mud layers having different densities collected from one location of Port of Hamburg and pre-sheared at $100\ s^{-1}$ using Couette geometry. Inset shows the equilibrium structural parameter (G_∞'/G_o') for different mud layers. Reprinted from Ref. [51].

human activities which are also responsible for the existence of these layers. This organic matter can or cannot be mineral-associated organic matter (i.e., organic matter adsorbed at the mineral surface or trapped inside the particle) [56]. There are two common sources of organic matter in mud: (i) natural and (ii) anthropogenic. The natural sources include erosion of terrestrial topsoils, plant litter, planktonic and pelagic biomass while surface runoff and sewage waste contribute to the anthropogenic source of organic matter [32].

The existence of organic matter in mud is also known to significantly influence the rheological and cohesive properties of mud [10, 14, 15]. For instance, the rheological characteristics of mud have been investigated in literature by varying organic matter content and keeping density constant [16]. The results showed an increase in yield stresses and moduli of mud with increasing organic matter content, for a similar density value (see **Figure 10a**). However, further research is required to investigate the effect of type of organic matter/biopolymer at different pH or ionic concentrations on the rheological behavior of mud.

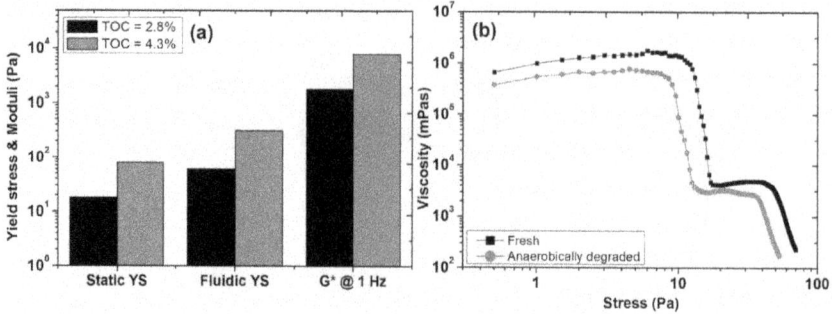

Figure 10.
(a) Yield stress values and complex modulus at 1 Hz for mud samples having similar density (1210 kg. M^{-3}) and different organic matter content obtained from port of Hamburg, adapted from ref. [16] (b) apparent viscosity as a function of shear stress for fresh and anaerobically degraded mud samples obtained from port of Hamburg.

In addition to organic matter content, its extent of degradation can also significantly affect the rheological properties of mud. Aerobic degradation (in the presence of oxygen) of organic matter usually results in the production of carbon dioxide while anaerobic degradation produces methane in addition to carbon dioxide [32]. For a detailed information about the aerobic and anaerobic degradation of mud, see ref. [32]. The entrapped gas bubbles of methane can significantly decrease the density and strength of mud, due to the poor solubility of methane in water. The outcome of stress ramp-up tests for fresh and anaerobically degraded mud samples, collected from port of Hamburg, is shown in **Figure 10b**. It can be seen that the values of the two yield stresses (static and fluidic) for degraded sample are significantly lower than the fresh ones. However, further quantification of organic matter content before and after degradation is required, in order to correlate the organic matter degradation with rheological characteristics of mud.

6. Discussion

The yield stress dependency on mud density is observed to vary for the samples collected from different parts of the world. As an example, fluidic yield stress values are plotted as a function of density for the samples collected from different ports (see **Figure 11**). One observes that the mud samples obtained from different ports exhibit considerably different yield stress values for a particular density. This difference may be attributed to the composition of mud, particle size distribution, type and content of TOC, ionic strength, etc. This behavior highlights the needs for a systematic investigation of the rheological properties of mud, as function of relevant parameters, for different ports.

Furthermore, the values of the rheological characteristics including yield stress and storage modulus of mud samples collected from different parts of the world are compared in **Table 3**. It can be seen that the mud from the Port of Santos [27], the Hangzhou Bay, China [23], the Port of Rotterdam [9], and the Port of Hamburg [16] display similar values of rheological properties for similar densities. However, the mud samples obtained from Mouth of Yangtze River, Shoal of Hangzhou Bay, and Yangcheng Lake, China [44] possess higher values of rheological parameters, which may be attributed to the their higher densities. Moreover, the mud from Eckernförde Bay, Germany show considerably lower yield stress values for the

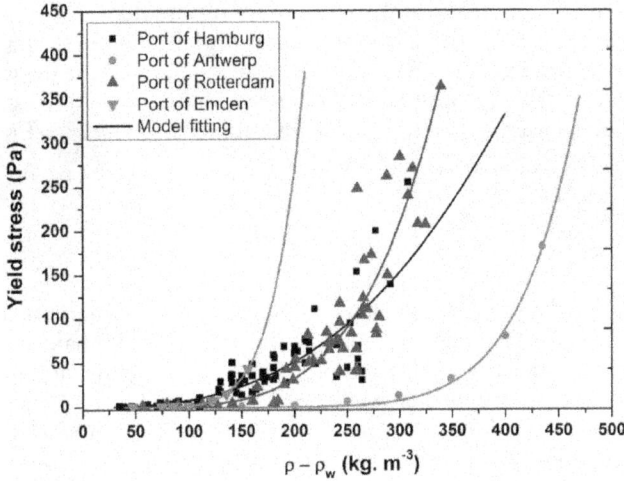

Figure 11.
Fluidic yield stress as a function of excess density $(\rho - \rho_w)$ for mud samples collected from different ports.

Location	Bulk Density (kg/m^3)	Fluidic Yield Stress (Pa)	Storage Modulus @ 1 Hz (Pa)	Ref.
Port of Rotterdam, the Netherlands	1168	7	45	[9]
Eckernförde Bay, Germany	1038–1280	1.07–20.50	—	[31]
Hangzhou Bay, China	1145–1634	0.55–40	0.02–15	[23]
Mouth of Yangtze River, China	1650–1700	910–2810	—	[44]
Shoal of Hangzhou Bay, China	1705–1741	772–2140	—	[44]
Yangcheng Lake, China	1651–1691	2070–3960	—	[44]
Lianyungang Port, China	1098–1305	0.098–28.029	2–1050	[29]
Port of Santos, Brazil	1085–1206	5–334	—	[27]
Port of Rio Grande, Brazil	1132–1308	5–350	—	[27]
Port of Itajaí, Brazil	1138–1360	5–299	—	[27]
Amazon South Channel	1293–1512	5–379	—	[27]
Port of Hamburg, Germany	1087–1210	2.44–312	0.47–7915	[16]

Table 3.
Comparison of rheological properties (i.e., yield stress and storage modulus) of mud samples obtained from different sources.

comparable densities, which may be linked to the organic matter content or measuring protocol.

As already mentioned that the yield stress can be used as a criterion to define navigable fluid mud layers, Port of Emden, Germany is reported to use the yield stress of 100 Pa as a criterion for nautical bottom [15]. However, this critical value of yield stress for defining nautical bottom is significantly dependent on the spatial variation of the sediment characteristics (i.e., sand content, organic matter content, etc.). For instance, the rheological properties of mud samples from Port of

Hamburg, Germany are significantly different for different locations within the port, due to the different organic matter content [46]. Therefore, using a single value of yield stress as a criterion for defining nautical bottom for the whole port can be quite tricky and instead, different boundaries of yield stress as a function of density can be used for different locations, in order to define the nautical bottom.

7. Conclusions

In this chapter the rheological behavior of mud as found in harbors, is discussed. Different mud layers, formed as a result of either natural or human activities, were defined. These mud layers exhibit a complex rheological fingerprint, by displaying a combination of thixotropy, two-step yielding behavior and viscoelasticity, which is conventionally associated to the existence of clay flocs (aggregated clay particles with organic matter). The analysis of the rheological properties of the top layer (fluid mud layer) is crucial for navigational purposes, optimizing dredging operations and the proper maintenance of dredged navigational channels.

In order to study the rheology of mud in laboratory, it was found that core sampling is the best sampling technique as it allows to collect mud samples without much disturbance. In contrast to what some authors do, it is not recommended to dilute a specific sample to predict its rheological behavior as function of density. It is shown that the rheological properties of natural mud layers of different densities found on top of each other at a specific location in the harbor do not match the properties of samples obtained from diluting the densest (deepest) mud layer sample. The reason lays in the differences in mud composition and structure at different depths.

The determination of yield stress of mud is highly dependent on the selected rheological geometry and experimental method. A detailed analysis shows that the Couette geometry along with stress ramp-up test is the most suitable combination to analyze the yield stress of mud for ports and waterways applications. The optimization of this stress ramp-up test enables to reduce the experimental time for different mud layers ($\sim$ 10–200 s). Several empirical or semi-empirical models are available in literature to fit the experimental data of mud displaying a single-step yielding. However, the mud samples are observed to exhibit a two-step yielding and, therefore, the behavior of shear stress as a function of shear rate (i.e., flow curve) can be represented as a sum of two functions, which capture the two yield regions. The model captures the two-step yielding phenomenon in mud samples quite well, within the density range of 1050–1200 kg. m^{-3}.

Several factors are known to influence the rheological characteristics of mud such as density and organic matter content. An exponential relation between yield stress and density (i.e., solid content) is usually observed in literature for mud from different sources. However, this correlation between yield stress and density is highly dependent on the mud composition. Apart from yield stress, other rheological properties including moduli, thixotropy, structural recovery, etc. are also strongly dependent on the density and composition of mud samples. For instance, the fluidic yield stress of mud from Port of Hamburg, Germany is observed to increase from 79 Pa to 312 Pa by increasing the organic matter content from 2.8% to 4.3%. The degradation of organic matter in mud, which can occur over time for different layers is found to significantly influence the rheological and cohesive properties of mud. Further research is required to investigate the effect of type of organic matter at different pH or ion type and concentrations on the rheological fingerprint of mud.

Acknowledgements

This study is funded by the Hamburg Port Authority and carried out within the framework of the MUDNET academic network: https://www.tudelft.nl/mudnet/

Conflict of interest

The authors declare no conflict of interest.

Author details

Ahmad Shakeel[1,2*], Alex Kirichek[3,4] and Claire Chassagne[1]

1 Department of Hydraulic Engineering, Faculty of Civil Engineering and Geosciences, Delft University of Technology, Stevinweg, The Netherlands

2 Department of Chemical, Polymer and Composite Materials Engineering, University of Engineering and Technology (New Campus), Lahore, Pakistan

3 Faculty of Civil Engineering and Geosciences, Delft University of Technology, The Netherlands

4 Deltares, The Netherlands

*Address all correspondence to: a.shakeel@tudelft.nl

IntechOpen

References

[1] Mehta AJ. An introduction to hydraulics of fine sediment transport: World Scientific Publishing Company; 2013.

[2] Ross MA, Mehta AJ. On the Mechanics of Lutoclines and Fluid Mud. Journal of Coastal Research. 1989:51-62.

[3] Gordon RB. Dispersion of dredge spoil dumped in near-shore waters. Estuarine and Coastal Marine Science. 1974;2(4):349-358. https://doi.org/10.1016/0302-3524(74)90004-8

[4] Harrison W, Wass ML. Frequencies of infaunal invertebrates related to water content of Chesapeake Bay sediments. Southeastern Geology. 1965; 6(4):177-186.

[5] Inglis C, Allen F. The regimen of the thames estuary as affected by currents, salinities, and river flow. Proceedings of the Institution of Civil Engineers. 1957;7 (4):827-868. https://doi.org/10.1680/iicep.1957.2705

[6] McAnally WH, Friedrichs C, Hamilton D, Hayter E, Shrestha P, Rodriguez H, et al. Management of fluid mud in estuaries, bays, and lakes. I: Present state of understanding on character and behavior. Journal of Hydraulic Engineering. 2007;133(1):9-22. https://doi.org/10.1061/(ASCE)0733-9429(2007)133:1(9)

[7] Whitehouse R, Soulsby R, Roberts W, Mitchener H. Dynamics of Estuarine Muds: A Manual for Practical Applications: Thomas Telford; 2000.

[8] Coussot P. Mudflow Rheology and Dynamics. Rotterdam: CRC Press; 1997. 272 p.

[9] Van Kessel T, Blom C. Rheology of cohesive sediments: comparison between a natural and an artificial mud. Journal of Hydraulic Research. 1998;36 (4):591-612. https://doi.org/10.1080/00221689809498611

[10] Malarkey J, Baas JH, Hope JA, Aspden RJ, Parsons DR, Peakall J, Paterson DM, Schindler RJ, Ye L, Lichtman ID, Bass SJ, Davies AG, Manning AJ, Thorne PD. The pervasive role of biological cohesion in bedform development. Nature Communications. 2015;6:6257. https://doi.org/10.1038/ncomms7257

[11] Parsons DR, Schindler RJ, Hope JA, Malarkey J, Baas JH, Peakall J, Manning AJ, Ye L, Simmons S, Paterson DM, Aspden RJ, Bass SJ, Davies AG, Lichtman ID, Thorne PD. The role of biophysical cohesion on subaqueous bed form size. Geophys Res Lett. 2016;43 (4):1566-1573. https://doi.org/10.1002/2016GL067667

[12] Paterson DM, Crawford RM, Little C. Subaerial exposure and changes in the stability of intertidal estuarine sediments. Estuarine, Coastal and Shelf Science. 1990;30(6):541-556. https://doi.org/10.1016/0272-7714(90)90091-5

[13] Paterson DM, Hagerthey SE. Microphytobenthos in Constrasting Coastal Ecosystems: Biology and Dynamics. In: Reise K, editor. Ecological Comparisons of Sedimentary Shores. Berlin, Heidelberg: Springer Berlin Heidelberg; 2001. p. 105-125.

[14] Schindler RJ, Parsons DR, Ye L, Hope JA, Baas JH, Peakall J, Manning AJ, Aspden RJ, Malarkey J, Simmons S, Paterson DM, Lichtman ID, Davies AG, Thorne PD, Bass SJ. Sticky stuff: Redefining bedform prediction in modern and ancient environments. Geology. 2015;43(5):399-402. https://doi.org/10.1130/G36262.1

[15] Wurpts R, Torn P. 15 years experience with fluid mud: Definition of the nautical bottom with rheological

parameters. Terra et Aqua. 2005; 99:22-32.

[16] Shakeel A, Kirichek A, Chassagne C. Is density enough to predict the rheology of natural sediments? Geo-Marine Letters. 2019;39(5):427-434. DOI: 10.1007/s00367-019-00601-2

[17] Shakeel A, Kirichek A, Chassagne C. Rheological analysis of mud from Port of Hamburg, Germany. Journal of Soils and Sediments. 2020;20:2553-2562. 10.1007/s11368-019-02448-7

[18] Richard Whitehouse RS, William Roberts, Helen Mitchener. Dynamics of estuarine muds: A manual for practical applications: Thomas Telford; 2000.

[19] Parker WR, Kirby R. Time dependent properties of cohesive sediment relevant to sedimentation management-European experience. Estuarine Comparisons: Academic Press; 1982. p. 573-589.

[20] May EB. Environmental effects of hydraulic dredging in estuaries: Alabama Marine Resources Laboratory; 1973. 88 p.

[21] Kirichek A, C Chassagne, H Winterwerp, Vellinga T. How navigable are fluid mud layers? Terra et Aqua. 2018;151:6-18.

[22] Kirichek A, Rutgers R, Nipius K, Ohle N, Meijer H, Ties T, et al. Current surveying strategies in ports with fluid mud layers. Hydro18; Sydney, Australia 2018.

[23] Huang Z, Aode H. A laboratory study of rheological properties of mudflows in Hangzhou Bay, China. International Journal of Sediment Research. 2009;24(4):410-424. https:// doi.org/10.1016/S1001-6279(10)60014-5

[24] Shakeel A, Kirichek A, Chassagne C. Rheological analysis of natural and diluted mud suspensions. Journal of

Non-Newtonian Fluid Mechanics. 2020; 286:104434.

[25] Coussot P. Mudflow rheology and dynamics: Routledge; 2017.

[26] Carneiro JC, Fonseca DL, Vinzon SB, Gallo MN. Strategies for Measuring Fluid Mud Layers and Their Rheological Properties in Ports. Journal of Waterway, Port, Coastal, and Ocean Engineering. 2017;143(4):04017008. https://doi.org/10.1061/(ASCE) WW.1943-5460.0000396

[27] Fonseca DL, Marroig PC, Carneiro JC, Gallo MN, Vinzón SB. Assessing rheological properties of fluid mud samples through tuning fork data. Ocean Dynamics. 2019;69(1):51-57. https://doi.org/10.1007/s10236-018-1226-9

[28] Yang W, Yu G. Rheological Response of Natural Soft Coastal Mud under Oscillatory Shear Loadings. Journal of Waterway, Port, Coastal, and Ocean Engineering. 2018;144(4): 05018005. doi:10.1061/(ASCE) WW.1943-5460.0000461

[29] Xu J, Huhe A. Rheological study of mudflows at Lianyungang in China. International Journal of Sediment Research. 2016;31(1):71-78. https://doi. org/10.1016/j.ijsrc.2014.06.002

[30] Ibanez Sanz M. Flocculation and consolidation of cohesive sediments under the influence of coagulant and flocculant [PhD Thesis]: Delft University of Technology; 2018.

[31] Fass RW, Wartel SI. Rheological properties of sediment suspensions from Eckernforde and Kieler Forde Bays, Western Baltic Sea. International Journal of Sediment Research. 2006;21 (1):24-41.

[32] Zander F, Heimovaara T, Gebert J. Spatial variability of organic matter degradability in tidal Elbe sediments.

Journal of Soils and Sediments. 2020;20(6):2573-2587. 10.1007/s11368-020-02569-4

[33] McAnally WH, Teeter A, Schoellhamer D, Friedrichs C, Hamilton D, Hayter E, et al. Management of Fluid Mud in Estuaries, Bays, and Lakes. II: Measurement, Modeling, and Management. Journal of Hydraulic Engineering. 2007;133(1):23-38. doi: 10.1061/(ASCE)0733-9429(2007)133:1(23)

[34] James A, Williams D, Williams P. Direct measurement of static yield properties of cohesive suspensions. Rheologica Acta. 1987;26(5):437-446.

[35] Nguyen QD, Akroyd T, De Kee DC, Zhu L. Yield stress measurements in suspensions: an inter-laboratory study. Korea-Australia Rheology Journal. 2006; 18(1):15-24.

[36] Zhu L, Sun N, Papadopoulos K, De Kee D. A slotted plate device for measuring static yield stress. Journal of Rheology. 2001;45(5):1105-1122.

[37] Steffe JF. Rheological methods in food process engineering: Freeman press; 1996.

[38] Uhlherr P, Guo J, Tiu C, Zhang X-M, Zhou J-Q, Fang T-N. The shear-induced solid–liquid transition in yield stress materials with chemically different structures. Journal of Non-Newtonian Fluid Mechanics. 2005;125 (2-3):101-119.

[39] Cheng DC. Yield stress: a time-dependent property and how to measure it. Rheologica Acta. 1986;25(5): 542-554.

[40] Shakeel A, Kirichek A, Chassagne C. Yield stress measurements of mud sediments using different rheological methods and geometries: An evidence of two-step yielding. Marine Geology.

2020;427:106247. https://doi.org/10.1016/j.margeo.2020.106247

[41] Nie S, Jiang Q, Cui L, Zhang C. Investigation on solid-liquid transition of soft mud under steady and oscillatory shear loads. Sedimentary Geology. 2020; 397:105570. https://doi.org/10.1016/j.sedgeo.2019.105570

[42] Nosrati A, Addai-Mensah J, Skinner W. Rheology of aging aqueous muscovite clay dispersions. Chemical engineering science. 2011;66(2):119-127. https://doi.org/10.1016/j.ces.2010.06.028

[43] Shakeel A, MacIver MR, van Kan PJM, Kirichek A, Chassagne C. A rheological and microstructural study of two-step yielding in mud samples from a port area. Colloids and Surfaces A: Physicochemical and Engineering Aspects. 2021.

[44] Yang W, Yu G-l, Tan Sk, Wang H-k. Rheological properties of dense natural cohesive sediments subject to shear loadings. International Journal of Sediment Research. 2014;29(4):454-470. https://doi.org/10.1016/S1001-6279(14)60059-7

[45] Claeys S, Staelens P, Vanlede J, Heredia M, Van Hoestenberghe T, Van Oyen T, et al. A rheological lab measurement protocol for cohesive sediment. INTERCOH2015 Book of Abstracts. 2015.

[46] Shakeel A, Kirichek A, Talmon A, Chassagne C. Rheological analysis and rheological modelling of mud sediments: what is the best protocol for maintenance of ports and waterways? Estuarine, Coastal and Shelf Science. 2021.

[47] Toorman EA. An analytical solution for the velocity and shear rate distribution of non-ideal Bingham fluids in concentric cylinder viscometers. Rheologica Acta. 1994;33(3):193-202.

[48] Toorman EA. Modelling the thixotropic behaviour of dense cohesive sediment suspensions. Rheologica Acta. 1997;36(1):56-65.

[49] Mewis J. Thixotropy - a general review. Journal of Non-Newtonian Fluid Mechanics. 1979;6(1):1-20. https://doi.org/10.1016/0377-0257(79)87001-9

[50] Barnes HA. Thixotropy—a review. Journal of Non-Newtonian fluid mechanics. 1997;70(1-2):1-33. https://doi.org/10.1016/S0377-0257(97)00004-9

[51] Shakeel A, Kirichek A, Chassagne C. Effect of pre-shearing on the steady and dynamic rheological properties of mud sediments. Marine and Petroleum Geology. 2020;116:104338. https://doi.org/10.1016/j.marpetgeo.2020.104338

[52] Bingham EC. Fluidity and plasticity: McGraw-Hill; 1922.

[53] Herschel WH, Bulkley R. Proc. 29th Ann. 1926;26:621-630.

[54] Worrall W, Tuliani S. Viscosity changes during the ageing of clay-water suspensions. Trans Brit Ceramic Soc. 1964;63:167-185.

[55] Soltanpour M, Samsami F. A comparative study on the rheology and wave dissipation of kaolinite and natural Hendijan Coast mud, the Persian Gulf. Ocean Dynamics. 2011;61(2):295-309. https://doi.org/10.1007/s10236-011-0378-7

[56] Hedges JI, Keil RG. Organic geochemical perspectives on estuarine processes: sorption reactions and consequences. Marine Chemistry. 1999;65(1):55-65. https://doi.org/10.1016/S0304-4203(99)00010-9

Chapter 5

Flocculation in Estuaries: Modeling, Laboratory and In-situ Studies

Claire Chassagne, Zeinab Safar, Zhirui Deng, Qing He and Andrew J. Manning

Abstract

Modelling the flocculation of particles in a natural environment like an estuary is a challenging task owing to the complex particle-particle and particle-hydrodynamic interactions involved. In this chapter a summary is given of recent laboratory and in-situ studies regarding flocculation. A flocculation model is presented and the way to implement it in an existing sediment transport model is discussed. The model ought to be parametrized, which can be done by performing laboratory experiments which are reviewed. It is found, both from laboratory and in-situ studies, that flocculation between mineral sediment and organic matter is the dominant form of flocculation in estuarine systems. Mineral sediment in the water column is < 20 μm in size and its settling velocity is in the range [0–0.5] mm/s. Flocs can then be categorized in two types: flocs of size [20–200] μm and flocs of size > 200 μm. The origin of these two types is discussed. The two types of flocs are found at different positions in the water column and both have settling velocities in the range [0.5–10] mm/s.

Keywords: Mud, flocculation, sediment transport, population balance equation, Rhine ROFI, Yangtse, floc, aggregation, LISST, monitoring, logistic growth

1. Introduction

Numerical fine sediment transport models make use of hydrodynamic models to estimate the transport (advection and diffusion) of suspended matter in the water column. In most numerical models, a few classes of suspended matter are defined. Each class is defined as a collection of particles having the same (often time-invariant) settling velocity and a concentration of suspended matter per class (suspended mass per unit of volume). The models are calibrated using in-situ observations, whereby suspended mass concentrations are measured at given locations in time. The settling velocity chosen for each class is based on in-situ observation of settling velocity and model calibration. To give an order of magnitude, it is generally found that using 3 classes of particles, with settling velocities of the order of ≤ 0.01 mm/s, 0.1 mm/s and ≥ 1 mm/s enables to correctly predict the Suspended Particulate Matter (SPM) in space and time for a large number of situations in coastal areas [1–5].

In the context of sediment transport modeling, several open questions however remain.

The hypothesis that the classes of particles do not interact is for instance questionable in estuarine regions, where fine particles are known to be in the form of flocs. Several studies over the years have therefore concentrated on implementing flocculation in sediment transport models [6–9]. Flocs are aggregates of mineral sediment particles, most often combined with organic matter. The underlying question, in terms of (numerical) modeling is related to flocculation dynamics. Are the models used at present, based on Population Balance Equations (PBE) adequate to capture the physical processed occurring in-situ? Which alternative equations, representing the flocculation process, should otherwise be implemented in a numerical model? This question will be addressed in Section 2.

Modeling flocculation requires to know the relevant parameters that plays a role in the process. Some of these parameters are for instance salinity, shear stress and type of organic matter present. The influence of each parameter on flocculation can be studied in a systematic way in the laboratory, but how do lab studies relate to in situ measurements? How should these parameters be accounted for in a numerical model? This question is addressed in Section 3.

Settling velocities are difficult to assess in situ [10, 11]. Measurements of the settling velocities of particles in quiescent water can be done by performing on board experiments [12–14]. These experiments consist in carefully pipetting a sample and let particles settle in a column filled with water of same composition as the one at the sample location. The settling velocity of particles in still water is then recorded by video microscopy. From the videos, the size, aspect ratio and Stokes settling velocity of each particle can be estimated.

The disadvantages of this method are: (1) only a limited number of samples can be taken and (2) the structure and velocity of the flocs can be altered through sampling and during settling, due to collective particle effects. Especially point (1) is of concern. As sediment transport models are run over large period of times, the interaction between particles and hydrodynamics are better understood if longer time series of measurements can be performed. The longer time series measurements performed in situ enable to assess particle size distributions (PSD), the volume concentration of suspended particles and suspended sediment (mass) concentration (SSC) based on light scattering (and also acoustic) techniques. Combining these measurements, a rough estimate of the mean settling velocity of particles can be given, using Stokes law [15, 16]. The question is whether this mean settling velocity is in agreement with the on board settling experiments. This question is specifically discussed in Section 3.2. A brief summary and outlook is given in the last section.

2. Flocculation models

Traditionally, flocculation is modeled using population balance equations (PBE), which were introduced in 1917 by Smoluchowski [17–19]. These equations represent the change in concentration of classes of particles over time, whereby a class is defined as a collection of identical particles. Each particle (floc) in class k has the same size L_k, and contains the same number k of primary mineral clay particles. If n_k is the number of class k particles per unit of volume, there are $k \times n_k$ primary particles per unit of volume in class k. The change in time of the number of particles within a class $n_k(t)$ is a function of the collision frequency and the collision efficiency between particles, as well as a function of the break-up of an aggregate, usually due to shear.

PBE models have successfully been applied to model the flocculation of suspensions destabilized by addition of salt [18, 19]. An example is for instance the aggregation that is likely to occur when fine mineral sediment particles are advected from fresh to saline environment.

In the presence of organic matter, however, the flocculation mechanisms cannot properly be modeled using PBE's for the following reasons:

- the size of a floc is not connected anymore to the number of primary mineral clay particles that composes the floc, since for a same size, a floc could be formed by aggregation of different amounts of organic matter and mineral sediment. Particles can break-up due to shear, but polymeric flocs are elastic and usually tend to coil under shear without breaking. Their shape and size thus can change over time from elongated to spherical without loss of mass.

- the collision frequency in PBE models is a function of the diameter of the colliding particles. Organic particles can have very anisotropic shape, and extracellular polymeric substances (EPS) that are a major driver for flocculation consist of elongated, flexible polymeric chains, composed mainly of polysaccharides, proteins and DNA. Their radius of gyration is a function of shear and water properties (such as salinity). The expression for the collision frequency is in this case unknown.

Some authors have tried to adapt PBE models to mimic the floc size evolution, but doing this implies to add a significant amount of unknown parameters to the model, even for a model accounting for only 3 classes of particles (microflocs, macroflocs and megaflocs) [6]. To parametrize the model, the following adjustable parameters are required (number of parameters in parenthesis): the mass fraction of microflocs produced when a macrofloc or a megafloc breaks up (2); the mass fraction of the remaining megafloc when a larger megafloc breaks up (1); the number of generated microflocs and macroflocs when a larger macrofloc or megafloc breaks up (4); the number of microflocs in one macrofloc or megafloc, or fractal dimension of microflocs and macroflocs (2); the collision efficiency, taken to be constant, but could be class-dependent (1); the collision frequency between microflocs, macroflocs and megaflocs (6); the breakup frequency of a megafloc and a macrofloc (2). These 18 parameters are difficult to estimate and therefore they are used as calibration parameters.

Recently, a simpler approach to model flocculation, that makes use of logistic growth theory, was proposed [20]. Logistic growth models are conveniently used to model systems whereby rate constants can be measured, such as the growth and decay of a bacterial community over time. In the context of flocculation, one can think of increase and decrease of the number of particles within a size class in terms of growth (birth) and decay. The time evolution of the concentration of particles within a class n (we here omit the subscript k for simplicity) is given by:

$$\frac{dn}{dt} = [b(t) - d(t)]n \tag{1}$$

where the birth function $b(t)$ and the decay function $d(t)$ are given by

$$b(t) = \frac{1}{t_b} \cdot \frac{a_b \exp\left(-\frac{t}{t_b}\right)}{1 + a_b \exp\left(-\frac{t}{t_b}\right)} \tag{2}$$

$$d(t) = \frac{1}{t_d} \cdot \frac{a_d \exp\left(-\frac{t}{t_d}\right)}{1 + a_d \exp\left(-\frac{t}{t_d}\right)} \tag{3}$$

There are 4 unknown parameters, a_d, a_b, t_d and t_b for each class of particles. Birth and decay are associated with the characteristic timescales t_b and t_d and a_b and a_d are parameters that influences the flocculation rates, see Eq. (7). The analytical solution of Eq. (1) is given by

$$n(t) = n_\infty \frac{1 + a_d \exp\left(-\frac{t}{t_d}\right)}{1 + a_b \exp\left(-\frac{t}{t_b}\right)} \tag{4}$$

The parameter n_∞ represents the value of $n(t)$ at long times, after the particles might have experienced birth and decay (or only birth or only decay). The flocculation rate dn/dt can be defined as being the slope of $n(t)$ at the onset of aggregation,

$$n(t \to 0) = n(t = 0) + \frac{dn}{dt} \times t \tag{5}$$

with

$$n(t = 0) = \frac{1 + a_d}{1 + a_b} n_\infty \tag{6}$$

$$\frac{dn}{dt} = n_\infty \frac{a_b(1 + a_d)/t_b - a_d(1 + a_b)/t_d}{(1 + a_b)^2} \tag{7}$$

The main advantage of Eqs. (1) and (4) is that each class can be seen as independent of each other: it is possible to estimate the evolution of one class only (for instance the class corresponding to the most abundant type of particles found in the water column). With 4 parameters, it is hence possible to parametrize the flocculation kinetics through Eq. (1) which is the equation required in numerical models, see Section 2.1.

2.1 Classes of particles

In the previous section, a new model was proposed to study the time evolution of a class of particles. As discussed in that section, because of the presence of organic matter, a class of particles cannot be defined as flocs containing the same number of primary particles. This is why, for the model proposed in Section 2.2, two types of particles will be distinguished: "primary" particles (Class 1) which are unflocculated mineral sediment particles and "flocs" (Class 2), which are particles aggregated with an unspecified amount of organic matter. The settling velocity of primary particles can be assumed to be a constant, but the settling velocity of flocs is a function of time, as the floc can get denser under the action of shear, or gain in volume and mass by further aggregation.

Class 1: class of particles defined as being mineral sediment particles. The mass concentration of Class 1 in the water column is $m_1(t)$ and the settling velocity associated to this class is $w_{s,1}$ which is assumed to be constant. The total clay mass in Class 1 per unit of volume is given by $m_1(t) = V_1(t) \times \rho_p$ where V_1 is the total volume occupied by Class 1 particle per unit of volume and ρ_p the absolute density of clay particles, which is of the order of 2600 kg/m^3. The total volume of particles in Class 1 per unit of volume is given by $V_1 = n_1(t) \times V_p$ where V_p is the volume of a particle in Class 1. In case that Class 1 is composed of polydisperse particles, one can subdivide Class 1 in different fractions based on size: $V_1 = \sum n_{1,i}(t) \times V_{p,i}$

where each particle in sub-class i has a volume $V_{p,i}$. The volumes V_1 and V_p can be estimated from in-situ particle size measurements, see Section 3.

Class 2: class of particles defined as flocs. The mass $m_2(t)$ represents the mineral clay mass (not the floc mass) per unit volume contained in Class 2. The settling velocity associated to flocs in this class is $w_{s,2}(t)$ which is assumed to be time-dependent. We will see, from in-situ measurements, that Class 2 can be split in two (Class 2a and Class 2b). Class 2b flocs have a smaller density and larger size than Class 2a flocs but a comparable settling velocity range.

Two types of flocculation are distinguished:

Microflocculation: this process describes the capture of primary particles by suspended organic matter or by existing flocs. This implies a transfer of mineral sediment mass between Class 1 and Class 2. A primary particle that is captured will experience a change in settling velocity, and be transported differently. A small colloidal mineral particle can be transported over larger distances than the same particle when it is imbedded in a floc. The sources and sink terms of mineral clay in the water column are located at the boundaries of the domain: mineral clay can be advected from the rivers into the sea or re-suspended from the bed due to shear. The resuspension of unflocculated mineral sediment from the bed occurs:

- during storm or dredging periods, when the fluff layer (containing organic matter) that constitutes the top of the sediment bed has been eroded.

- when the organic matter has decayed sufficiently to release primary particles from the bed. This happens mainly during the winter season.

Macroflocculation: this process describes the capture of a floc by another floc. There is no mass transfer between classes as all particles experiencing macroflocculation remain in Class 2. We will see that this process, even though occurring in the water column, is most probably not the dominant one in terms of sediment transport. When two flocs aggregate, their settling velocity will change. However, we will show that there is a large spread in settling velocity for flocs, even without flocculation, as the settling velocity changes due to coiling under shear. For numerical modeling purposes, it is therefore not necessary to account for macroflocculation, as no clear correlation can be made, at this stage, between macroflocculation and change in settling velocity.

Flocculation, in terms of numerical modeling, is hence defined as the mineral sediment mass transfer between Class 1 and Class 2. All the primary particles which leave Class 1 by flocculation become part of Class 2. Flocs that aggregate or break in smaller flocs remain part of Class 2. The transfer from Class 2 to Class 1 occur when organic matter in flocs has decayed sufficiently to free mineral particles.

2.2 Inclusion of flocculation in a sediment transport model

The inclusion of flocculation in a sediment transport model is done by expressing the advection–diffusion equations for the two mass classes of particles. The equations presented below are for the special case where only vertical advection and diffusion is considered. Generalization to other coordinates is straightforward.

Step 1 At the first numerical step, particles are entering or leaving a volume element by advection/diffusion:

$$\frac{\partial m_1}{\partial t} + \frac{\partial((v_z - w_{s,1})m_1)}{\partial z} - \frac{\partial}{\partial z}\left(D_z \frac{\partial m_1}{\partial z}\right) = 0 \tag{8}$$

$$\frac{\partial m_2}{\partial t} + \frac{\partial((v_z - w_{s,2})m_2)}{\partial z} - \frac{\partial}{\partial z}\left(D_z \frac{\partial m_2}{\partial z}\right) = 0 \tag{9}$$

The parameters v_z and D_z represent the vertical water velocity and eddy diffusion and w_s represents the vertical settling velocity of a particle under the influence of gravity. The new mass concentrations $m_1(t + dt/2)$ and $m_2(t + dt/2)$ are obtained.

Step 2 At the second numerical step, flocculation occurs within the volume element, and a mass transfer occurs between Class 1 and 2:

$$\frac{\partial m_2}{\partial t} = -\frac{\partial m_1}{\partial t} \tag{10}$$

The mass transfer can be modeled by equations similar to Eq. (1). In the simple case where only aggregation (no break-up) occurs, one gets

$$\frac{\partial m_1}{\partial t} = -d_1(t) \times m_1 \tag{11}$$

$$\frac{\partial m_2}{\partial t} = b_2(t) \times m_2 \tag{12}$$

And it follows from Eq. (10) that $d_1(t) \times m_1 = b_2(t) \times m_2$. Both birth and decay functions are function of the mass of organic matter present in the water. The new mass concentrations $m_1(t + dt)$ and $m_2(t + dt)$ are obtained. The dynamics of the mass transfer between Class 1 and Class 2 are discussed in Section 3.

The average settling velocity of Class 1 particles is given by

$$w_{s,1} = \frac{d_p^2}{18\eta}\left(\rho_p - \rho_w\right)g \tag{13}$$

This velocity is a constant as function of time. The average size of primary particles can be assessed by Laser In-Situ Scattering and Transmissometry (LISST) data in-situ or laboratory PSD measurements from samples collected in-situ. Settling column experiments combined with video microscopy can confirm the values estimated for $w_{s,1}$.

Step 3 At the third numerical step, the change in settling velocity of Class 2 particles is addressed. The link between particle number, mass and volume concentration is not straightforward for Class 2 particles. Class 2 particles evolve in time, as flocs can aggregate, break or coil, hereby changing their size, aspect ratio and density, and therefore their settling velocity – a key parameter for numerical modeling. The settling velocity of Class 2 particles is given by

$$w_{s,2}(t) = \frac{[d_f(t)]^2}{18\eta}\left(\rho_f(t) - \rho_w\right)g \tag{14}$$

Whereby both the average size of flocs d_f and its density ρ_f should be updated as function of time. In Section 2.3 we describe how $w_{s,2}$ can be estimated from in-situ measurements and in Section 2.4 how analytical functions can be obtained from laboratory experiments.

Boundary condition The boundary condition at the fluid/bed interface can be written in terms of mass fluxes. One flux is the mass flux settling down to the bed and given by $w_{s,k}m_k$. The other is the erosion flux, which is usually written in the form $M_k(\tau/\tau_c - 1)$ where M_k is mass per unit of area and time that leaves the bed, τ is the bottom shear stress and τ_c is the stress at which the bed start to be eroded.

If the bed is composed of a fluff layer with underneath a organic matter-degraded bed, it is assumed that the erosion will be different for each layer. The mass transfer from Class 2 to Class 1 inside the bed by degradation of organic matter is given by an equation analogous to Eq. (10). The parametrization of this equation is an on-going topic of research.

2.3 Estimation of the settling velocity from in-situ measurements

To estimate the mean density of the flocs ρ_f, one assumed it is given by

$$\rho_f = \frac{m_f + m_w}{V_f} \tag{15}$$

where m_f(g) is the mass mineral sediment inside a floc and m_w(g) the mass water inside a floc (it is therefore assumed that the density of organic matter is close to the one of water) and V_f(L) is the volume of a floc. It follows that

$$\rho_f - \rho_w = \left(1 - \frac{\rho_w}{\rho_p}\right)\frac{m_f}{V_f} \tag{16}$$

where ρ_f is the floc density (g/L), ρ_w is the ambient water density (about 1000 g/L), ρ_p is the sediment absolute density (usually taken to be equal to the mineral sediment density, i.e. of the order of 2650 g/L). Realizing that

$$\frac{n_2 m_f}{n_2 V_f} = \frac{m_2}{V_2} \tag{17}$$

where V_2 (L/L) is the volume occupied by Class 2 flocs per unit of volume and assuming that $V_2 \gg V_1$ (where V_1 is the volume occupied by Class 1 particles per unit of volume) one gets

$$\rho_f - \rho_w = \left(1 - \frac{\rho_w}{\rho_p}\right)\frac{m_2}{V_{tot}} \tag{18}$$

where V_{tot}(L/L) is the total volume of particles detected per unit of volume. It represents the volume occupied by the sediment in a given volume of water and can be measured in-situ by LISST, which also provides full PSD's (in the range 2–500 μm) as function of time. The mean floc size d_f can therefore also be estimated from LISST data.

We recall that m_2(g/L) represents the total mass of mineral clay per unit volume inside Class 2. Note that most authors assume that $m_2 = m_{clay}$ (the total mass of mineral clay in suspension) hereby implying that there are no Class 1 particles in suspension. This approximation can lead to an overestimation of the floc density since despite representing a small volume of the total volume compared to Class 2, Class 1 particles may represent a non-negligible part of the total sediment mass. The mass m_{clay} is the mass suspended mineral sediment per unit of volume (g/L) also denoted SSC (suspended sediment concentration) and usually estimated in-situ by Optical Back Scattering (OBS) technique.

From the estimation of $\left(\rho_f - \rho_w\right)$ and d_f the settling velocity $w_{s,2}(t)$ can be evaluated.

2.4 Estimation of the settling velocity from models

Considering the fact that a lot of data has been collected over the years to link the mean floc size to parameters such as shear rate and salinity, the time evolution of the mean floc size d_f can be parametrized as function of these variables [21–25]. It has been shown for example that in the case of salt-induced flocculation the equilibrium floc size is given by

$$d_{f,eq} = CG^{-\gamma} \tag{19}$$

Where C and γ are constants to be fitted. Values γ are around [0.29–0.81] whereas for C they are in the range $[10^{-3} - 10^{-2}]$ m/s$^{1/2}$ [22]. As shown in Section 3.1.1, salt-induced grown flocs will never exceed the Kolmogorov microscale d_K given by

$$d_K = \left(\frac{G}{\nu}\right)^{-0.5} \tag{20}$$

Where ν is the kinematic viscosity which is of the order of 10^{-6} m^{2}s^{-1} for water at 20°C. From laboratory experiments, the time evolution of the mean particle size can be modeled using the same type of logistic growth model as presented in Eq. (4):

$$d_f(t) = d_{f,eq} \frac{1 + a_{f,d} \exp\left(-\frac{t}{t_{f,d}}\right)}{1 + a_{f,b} \exp\left(-\frac{t}{t_{f,b}}\right)} \tag{21}$$

where the parameters $d_{f,eq}$, $a_{f,d}$, $a_{f,b}$, $t_{f,d}$ and $t_{f,b}$ are found by fitting experimental results. By performing a large number of laboratory experiments, where each of the relevant parameters (salinity, organic matter, shear) can be varied independently of one another the dependence of $d_{f,eq}$, $a_{f,d}$, $a_{f,b}$, $t_{f,d}$ and $t_{f,b}$ on these parameters can be found. This work is currently going on [20]. The floc density is found from settling velocity measurements, and is usually parametrized using the relation

$$\rho_f - \rho_w = \left(\rho_p - \rho_w\right)\left(\frac{d_f}{d_p}\right)^{D-3} \tag{22}$$

where D is a parameter (often designated as "fractal dimension") between 1.5 and 3 and d_p a characteristic size, such that $d_p \leq d_f$. A large amount of data is available for the parameters D and d_p, but a systematic study of their dependence on the relevant parameters is still missing.

From Eqs. (14), (20) and (21) the settling velocity $w_{s,2}(t)$ can be evaluated.

3. Laboratory studies and in-situ monitoring

3.1 Laboratory studies

Laboratory experiments have the great advantage that the sample under investigation is in a closed volume, and that therefore the mineral clay mass is conserved during the experiment. This enables to estimate mass balances that are required for flocculation models.

3.1.1 Flocculation by salt and pH

In **Figure 1**, some examples are given of the mean particle size evolution for different salinities and pH. At low pH, the edges of the clay particles are positively charged, leading to Coulombic attraction between the negatively charged faces and the positively charged edges. Flocs will be created whereby the particles preferably arrange themselves in a so-called house of card structure. At pH > 7, clay particles are overall negatively charged and flocculation is driven by van der Waals attraction, when the Coulombic repulsion has been screened by sufficient addition of salt [25]. The main features of these type of flocculation mechanism are:

- the time to reach a steady-state mean floc size is of the order of hours

- the flocs produced are always smaller than the Kolmogorov microscale

As discussed in the previous section, this type of flocculation will not be the preferred mode of aggregation in estuarine systems. Flocs in these systems will in majority contain some proportion of organic matter. Organic matter-induced flocculation is very fast, especially in saline environment, where Coulombic repulsion between particles of same charge is neutralized.

3.1.2 Flocculation by organic matter

An example is given here of polymer-induced flocculation. For this example 0.7 g/L of river clay with 4.7 mg/L polymeric cationic flocculant was used. Typical example of cationic flocculant in the water column are polysaccharides. The flocculant to clay ratio is 6 mg/g. The optimal flocculant dose is defined as the flocculant to clay ratio which leads the fastest to the creation of large flocs. The optimal dose for flocculation with this cationic flocculant for the studied clay was found to be around 5 mg/g flocculant to clay ratio [26]. Another example can be found in [20], where a ratio of 0.71 mg/g was used (lower than optimal dose).

The composition of the clay used is predominantly quartz, calcite, anorthite and muscovite [27]. The flocculant, referenced ZETAG 7587, is composed of a copolymer of acrylamide and quaternary cationic monomer usually used for the conditioning of municipal and industrial substrates.

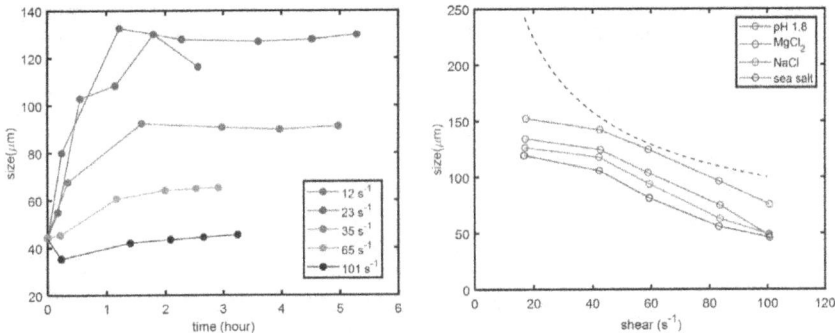

Figure 1.
Left panel: Time evolution of the mean particle of a kaolinite suspension (0.135 g/L) at pH = 9.3 with 100 mM of added $MgCl_2$ for different shear rates. Right panel: Mean particle size at steady-state as function of shear rate. Mud was taken from the lower Western Scheldt (0.135 g/L). The sample at pH = 1.8 has no added salt, the other samples were at pH = 8. The salts used (indicated in the legend) were: $MgCl_2$ (40 mM), NaCl (100 mM). The dashed line corresponds to the Kolmogorov microscale. Data is adapted from [18].

The particle size distribution (PSD) and mean particle size (D50) as function of time of this suspension was measured by static light scattering using a Malvern MasterSizer 2000, with a procedure described in [20, 27]. The PSD of the clay sample is given as the PSD at t = 0 s. At a time defined as t = 1 s flocculant was added to the clay suspension. The Particle Size Distributions (PSD) are given following the class distributions of the software of the static light scattering device. The size distribution is given by

$$d_k(\mu m) = 10^{0.05 \times k}/50 \qquad (23)$$

where k is an integer number between 1 and 100 and represents the number of the class associated to a given size (diameter) d_k. For instance, k = 46 corresponds to particle d_{46} = 4 μm and k = 84 corresponds to particle size d_{84} = 320 μm. A hundred size bins are so created. In the experiments the concentration of each size class is given in terms of percent volume concentration $V_{k,\%}$ (volume occupied by Class k particles divided by the volume occupied by all particles) and consequently $\sum_{k=1}^{100} V_{k,\%} = 100\%$.

The samples were further analyzed by video microscopy. This was done using a LabSFLOC-2 camera system (Laboratory Spectral Flocculation Characteristics, version 2) which records the settling velocity of particles from a pipetted amount of sample. From the settling velocity, the particle size, shape and density were estimated [28, 29].

The time evolution of the PSD of this suspension is given in **Figure 2**. The flocculation is very fast, as the cationic flocculant concentration is close to the optimal dosage.

One can see that the results obtained by video microscopy are in close agreement with the ones obtained from laser scattering. The % volume of particles below 100 μm is larger by video microscopy than by laser scattering. The PSD peak obtained by laser scattering is wider than the one obtained by video microscopy. It was observed that PSD's obtained from this particle sizer overestimated the largest sizes [27]. Consequently, as the total volume should give 100%, the % volume of smaller particles is underestimated.

Two types of fits were performed: first the data set was fitted considering times below 250 s, and then the same data set was fitted for the duration of the experiment (1835 s). The corresponding time evolution of the different size classes are given in **Figure 3**.

From the analysis of all the size classes, it is clear that some classes are not representative of a flocculation process, like class size 106 μm in **Figure 3**, which is

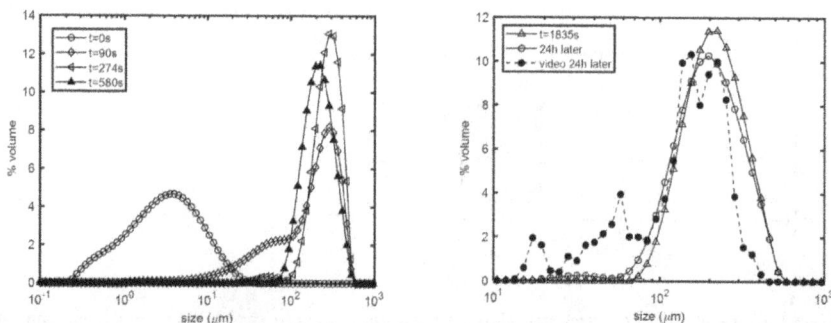

Figure 2.
Time evolution of the PSD of a clay suspension (0.7 g/L) in presence of 4.7 mg/L of cationic flocculant (added at t = 1 s); the video microscopy data has been acquired using the LabsFLOC-2 camera system.

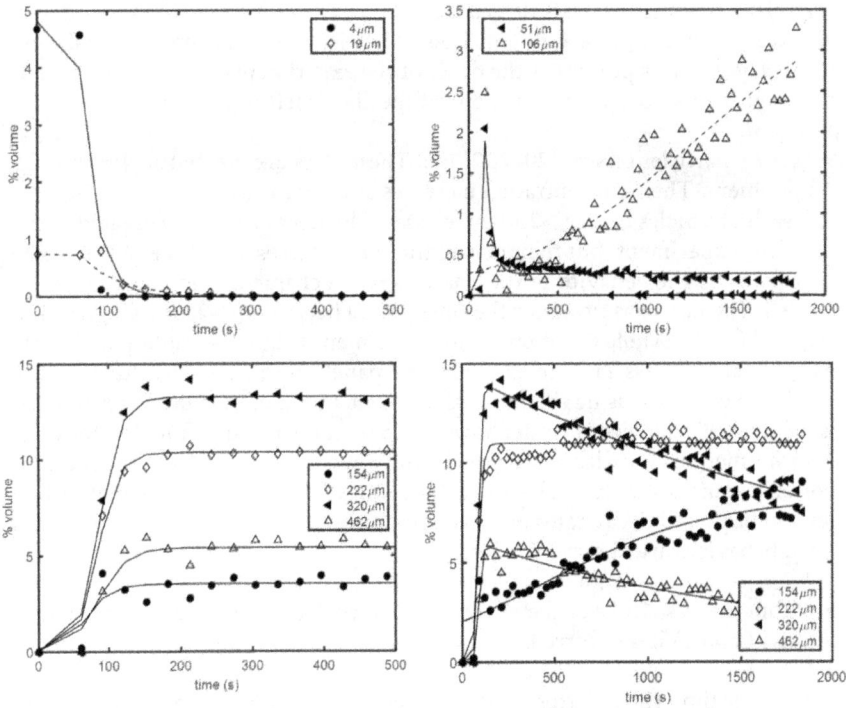

Figure 3.
Time evolution of the concentrations of a different size classes (given in the legends); clay suspension (0.7 g/L) in presence of 4.7 mg/L of cationic flocculant (added at t = 1 s). The lines represent fits obtained from the analytical model. Bottom panel, left: Fits for the period [0–500 s]. Bottom panel, right: Fits for the entire duration of the experiment.

located in-between the high-end tail of the initial clay PSD and the low-end tail of the flocculated clay PSD.

In order to represent the typical behavior of particles under flocculation, it is therefore better to define size classes as wider groups, containing particles in a given size range. From the fits of all classes, three size classes are proposed. Similar classes have been identified by other authors, from in-situ studies [30, 31]. Replotting the data by creating three size classes gives **Figure 4**.

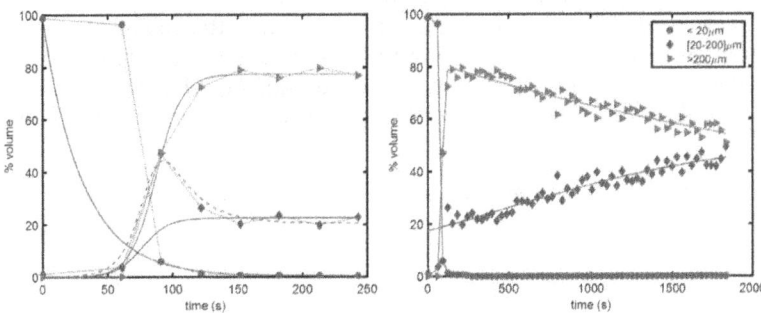

Figure 4.
Time evolution of the concentrations of three size classes (given in the legends); clay suspension (0.7 g/L) in presence of 4.7 mg/L of cationic flocculant (added at t = 1 s). The full and dashed lines represent fits obtained from the analytical model.

The characteristics of the 3 classes are:

Class 1: particles of size <20 µm. These particles represent the unflocculated mineral clay that was present at the onset of the experiments. The concentration of this class goes to zero over time as none of the clay is left unflocculated in this experiment.

Class 2a: particles of size [20–200] µm. These flocs are created at the onset of the experiment. Their concentration increases at longer times.

Class 2b: particles of size >200 µm classes. The flocs are largely created at the onset of the experiment, but their concentration decreases over time, due to coiling.

By analyzing the behavior of the classes, several characteristic times can be identified. The data was fitted for the time period between [0–250] s (**Figure 4**, left panel) and for the whole duration of the experiment (**Figure 4**, right panel). It is shown, by taking Class 2a as an example (left panel, short times fit) that the data can be fitted correctly using the full analytical equation Eq. (4) with birth and decay (dashed line). The choice is made, however to use only a decay function for Class 1 and birth functions for Classes 2a and 2b (full lines), so as to analyze the most important dynamics of these classes at short times. The associated characteristic times and birth and decay rates are given in **Table 1**.

Two behaviors are observed:

- within the first 100 s of the experiment, flocculation (mass transfer between Class 1 and 2) has occurred.

- at longer time (in a matter of hours) there is a significant change in size for Class 2 particles, whereby large flocs (from Class 2b) coil and start to populate size Class 2a.

Twenty-four hours after the start of the experiment, the sample was re-measured after the steering was stopped overnight. The found PSD is given in **Figure 2**, along the PSD found from the analysis of a subsample by video micros-copy. From video microscopy 1550 particles were recorded, and their sizes were divided into the same size classes as given by Eq. (23). The volume concentration of particles in each class was thus estimated. The settling velocity and estimated density of the particles from Stokes' law are given in **Figure 5**, along with the aspect ratio of each particle. One can observe that the aspect ratio is quite large for many particles. Most particles with high aspect ratio have been formed by differential settling during the video microscopy experiment, where it was observed that flocs that were touching immediately stuck to each other (a consequence of the fact that the flocculant to clay ratio is close to the optimal dose).

Using again Eq. (23) to create bin sizes, the data is replotted in **Figure 6**, top panel. The classes are furthermore divided in the three size classes 1, 2 and 3 (lower panel).

The density as function of size was estimated using Eq. (22). The characteristic size d_p was taken to be the smallest recorded particle, viz. 13 µm. The density ρ_p was

Class size	Flocculation rate (% s^{-1}) for [0–250] s	Flocculation rate (% s^{-1}) for [0–1835] s
Class 1: < 20 µm	−3.2 (t_d = 70 s)	0
Class 2a: [20–200] µm	0.0011 (t_b = 10 s)	0.018 (t_b = 673 s)
Class 2b: >200 µm	0.0012 (t_b = 10 s)	−0.022 (t_d = 2231 s)

Table 1.
Aggregation kinetics for the three classes. Characteristic times for birth or decay are given in parenthesis.

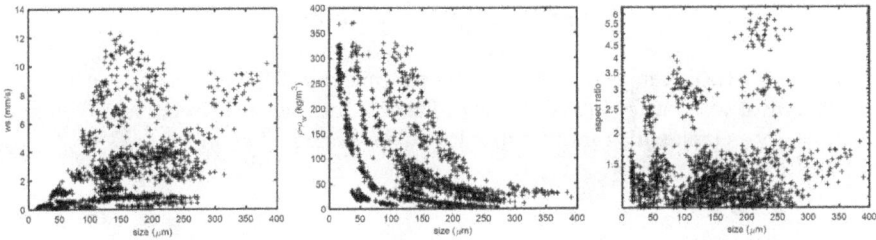

Figure 5.
Settling velocity, estimated density (from Stokes' law) and aspect ratio of the sample corresponding to "video 24h later" in **Figure 2.**

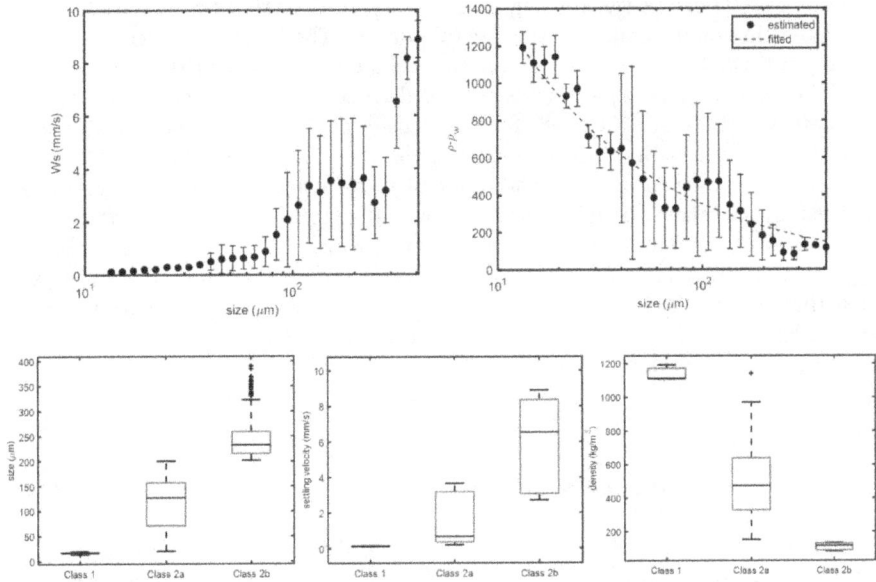

Figure 6.
Settling velocity, estimated density (from Stokes' law) and sizes of the sample corresponding to "video 24h later" in **Figure 2.**

taken to be the average density of the particles of class size 13 μm (2250 kg/m³), ρ_w is the density of water (1000 kg/m³), and D was found to be 2.39. It can be seen that, even if Eq. (22) is a good approximation for the density behavior, there is a large scatter in the measured data, as particles in Class 2a have a relative density that varies between 100 to 1000 kg/m³, resulting in settling velocities ranging between 0.5 and 5 mm/s.

For each PSD measurement using the laser diffraction technique the total volume of particles detected per unit of volume V_{tot} (L/L) is known. At t = 0 all clay is unflocculated and it is therefore easy to estimate the expected V_{tot} from the clay concentration in the jar (m_{clay}) and the absolute density of mineral sediment ρ_p:

$$V_{tot} = \frac{m_{clay}}{\rho_p} = \frac{0.7 \text{ g/L}}{2650 \text{ g/L}} = 0.0265\% \tag{24}$$

which is very close to the value of 0.0273% found by laser diffraction. In time, particles will aggregate, and mass will be transferred from Class 1 to Class 2. This mass transfer can be estimated by

$$m_1(t) = \rho_p V_1 = m_{clay} - m_2(t) \qquad (25)$$

Where V_1 (L/L) represents the volume occupied by Class 1 particles per unit of volume. Most software's (LISST, Malvern ParticleSizer) give the relative % volume occupied by a class, which implies that V_1 can be evaluated from

$$V_1 = V_{1,\%} V_{tot} \qquad (26)$$

Where $V_{1,\%}$ is the volume occupied by Class 1 particles divided by the volume occupied by all particles. When the system is unflocculated, $V_{1,\%} = 1$ and one recover Eq. (24).

The mass transfer is represented in **Figure** 7. It is clear that the change in mass as function of time can be fitted using the same logistic growth functions used to fit the change in volume concentration (see **Figure 4**). These functions can subsequently be implemented in the numerical model, see Eqs. (10)–(12).

The density of Class 2 particles is evaluated according to Eq. (18). It is found that that between 500 s and 2000s the relative density of Class 2 flocs increases linearly from 30 to 36 g/L. As was already evident from the PSD analysis, the flocs have become denser over time, under the action of shear. After 24 h and being re-suspended, the relative density became about 140 g/L, which has to be compared with the mean value found by video microscopy, which is 340 g/L. There has been a significant increase in density overnight. The effect of deposition/resuspension is a topic that needs to be investigated further. From the estimation of the change in density and mean floc size over time, the settling velocity $w_{s,2}(t)$ (see Eq. (14)) can be estimated and implemented in the numerical model.

3.2 In-situ studies

3.2.1 Observed size, shape and behavior under flow

The large spread in particle size, aspect ratio and settling velocity found in laboratory experiments was also observed during in-situ video recordings, see **Figures 8** and **9**, performed during a 13 hours survey in the Rhine Region Of

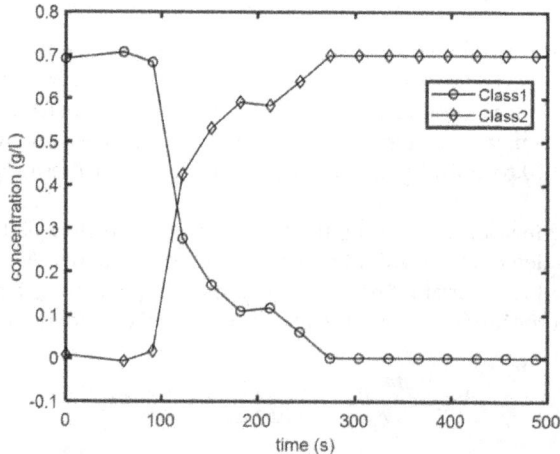

Figure 7.
Time evolution of the concentrations of mass classes 1 and 2 (given in the legends); clay suspension (0.7 g/L) in presence of 4.7 mg/L of cationic flocculant (added at t = 1 s).

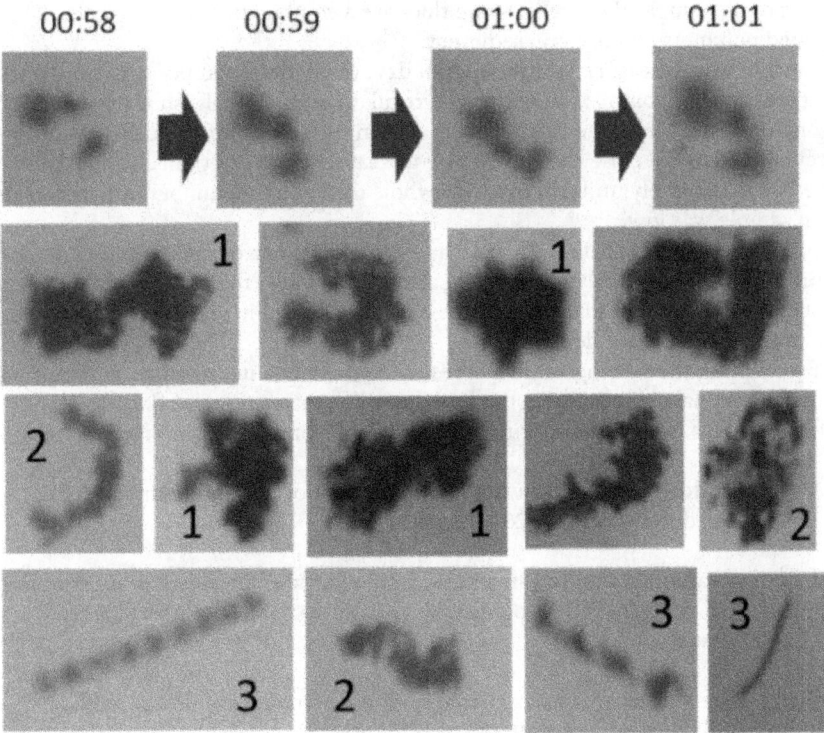

Figure 8.
Screenshots from the video recording taken with the underwater camera one meter above bed. Top panel:
Aggregation of two flocs (video time indicated above the picture). The flocs stuck together at 00:59 and
remained as one entity during the whole time they stayed in the field of view (until 01:02). Their shape adapted
to the flow, indicating an elastic behavior (see 01:00 and 01:01). Some screenshots of typical flocs of largest size
(100–500 μm) are given in the lower panels.

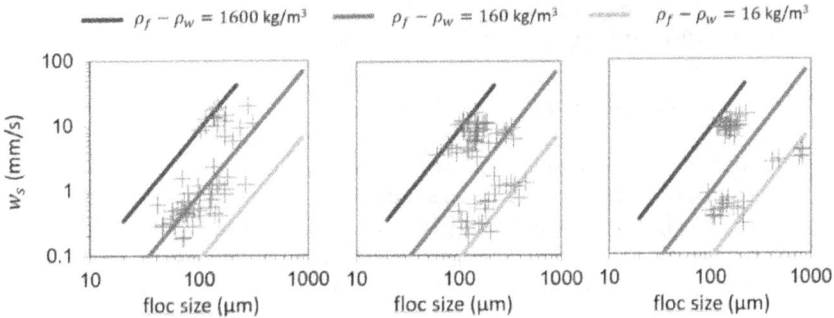

Figure 9.
Settling velocity distributions from LabsFLOC-2 measurements. From left to right, samples taken at: 06:00,
09:50 and 10:40 GMT. Data is adapted from [23].

Freshwater Influence (Rhine ROFI), about 10 km downstream of the mouth of the
Rotterdam waterway, during a calm weather day, with low shear stresses and low
SSC [32]. In **Figure 8** an illustration of in-situ flocculation is given: two flocs are
observed to stick to each other and remain stuck in the hydrodynamic flow,
displaying an elastic behavior. Flocs in situ are formed by aggregation of mineral
sediment and both living and dead organic matter. Living organic matter is illustrated by the elongated particles in the lower panel which are formed by

aggregation of single algae cells. Some flocs are a combination of living cells, excreted polymers and mineral sediment.

Despite being measured during a calm day, the hydrodynamics at the measurement location are complex, owing to the regular passage of a fresh water front originating from the Rhine river. This results in the advection and diffusion of flocculated material close the bed, where the camera was positioned. The PSD was nonetheless relatively uniform over the whole day, and the number of particles per taken sample was low.

In **Figure 9** three settling velocities measurements performed on samples taken at different times of the day using the LabsFLOC-2 video microscopy technique are given. One can observe that particles have a large spread in size and settling velocities (and hence relative density).

By coupling the settling velocities results to the video microscopy observations, the density of particles can be estimated and three major types of flocs could be distinguished, based on their structure (indicated by a number in **Figure 8**).

1. compact flocs, containing a significant amount of mineral sediment, with an estimated density close to 2600 kg/m^3

2. flocs of various shape and structure, from elongated to coiled, in most of them strains of algae are still recognizable, with an estimated density close to 1160 kg/m^3.

3. bare algae strains, or strains coated with little amounts of debris, with an estimated density close to 1016 kg/m^3.

As was already found in laboratory experiments, density and floc size cannot be properly correlated: there is a wide spread in density in the [20–200] μm size class (Class 2a). As a lot of flocs of larger size are anisotropic, their equivalent diameter will have them part of the Class 2a (and not 2b). Moreover aspect ratio cannot be a proper variable to estimate settling velocities, as flocs are elastic and are prone to coil over time. In the next subsection, we will introduce a better variable to distinguish between different types of settling particles.

3.2.2 Variation with depth, tidal cycle and season

Many studies have confirmed the role of bio-cohesion in the formation of flocs [30, 33, 34]. In two recent studies, it was found that there is a correlation between flocculation and algal microorganisms presence in the water column, also outside the algae bloom season [35, 36]. This correlation can be studied using the sediment to algae concentration ratio, which is expressed as

$$ratio = CC/SSC \tag{27}$$

where CC is the chlorophyll concentration (μg/L) and SSC the suspended sediment concentration (g/L). The data shown in **Figure 10** was collected in the South Passage of Changjiang Estuary (East China), for the summer period, when the amount of organic matter in the water column is significant [36]. A correlation was found between CC/SSC and particle density, evaluated from Eq. (18). The density was found to increase as function of CC/SSC in winter (when CC is constant over the whole water column), a low CC/SSC thus being associated with a high SSC. Particle density was found to decrease with CC/SSC in summer (when SSC is relatively constant over the whole water column), a low CC/SSC thus being

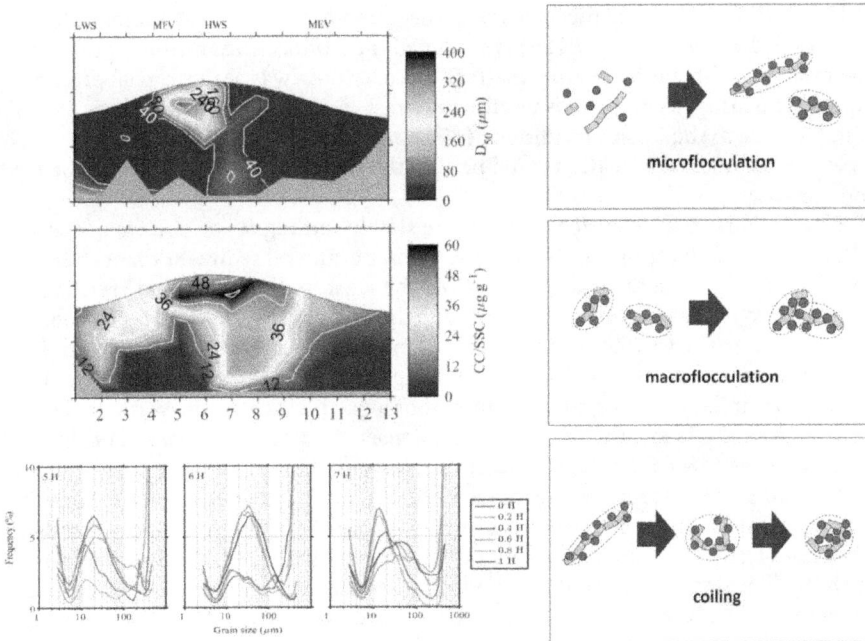

Figure 10.
Left panel: Recorded mean size (D50), estimated CC/SSC ratio and specific PSD's at different depths. Right panel: Schematic description of flocculation and re-sizing mechanisms happening in-situ. Data is taken from [28].

associated with a low CC. The D50 was found to be an exponentially decreasing function of particle density, in line with Eq. (22). Due the higher shear rates in summer, even silt particles could be entrapped in the flocs. These silt particles were even found at the top of the water column, as was assessed from laboratory PSD analysis on samples treated so as to remove organic matter.

The chlorophyll concentration (CC) was found to be relatively uniform over the whole water column in winter, with concentrations of about 0.6–1.0 µg/L. In summer the CC increased towards the bed with concentrations ranging from 1.3 to 3.7 µg/L. The CC/SSC ratio ranged from1.5 to 61.6 µg/L in summer, the higher values being found at the top of the water column, where algae activity is highest. A threshold value for CC/SSC was found to be 10–20 µg/L. Above the threshold value, flocs are predominantly governed by organic matter (algae), and bimodal PSD's are found, reflecting both the anisotropy of algae-containing flocs (known to lead to multimodal PSD peaks) along with their large spread in size. These flocs populate the whole water column in summer from the high water slack (HWS) period to ebb tide.

The PSD dynamics around HWS is particularly interesting and given in **Figure 10** as function of depth in the water column: 0H represents the surface water, whereas 1H represents the position just above bed. At 5 h, large particles are advected at the top of the water column from the seaside. At 6 h, two PSD's are observed. In the upper half of the water column [0H – 0.6H] the PSD peaks at about 20 µm (with a peak having an asymmetric shape towards the highest sizes – indicating the presence of large particles). In the lower half of the water column [0.6 – 1H] the PSD peaks at 30–50 µm and does not display any large PSD asymmetry. This transition seems to be in line with the change in salinity at HWS: algae-rich particles are trapped above the pycnocline, whereas clay-algae flocs are located underneath. This trapping mechanism has been reported by several authors. During the period that algae-rich particles are trapped, microflocculation (the capture of fine mineral sediment by organic matter) can occur. As was seen from laboratory experiments, this flocculation is usually very fast, of the order of seconds or minutes. Due to the in-situ conditions (different mixing, lower SSC), the timescale for microflocculation could be slightly different, but is expected to be fast nonetheless.

Subsequently the algae-rich particles are slowly settling to the bottom of the water column, still capturing the finest fraction of mineral sediment that can be found at any depth in the water column. As the algae-sediment flocs are settling they can also experience macroflocculation (the capture of a floc by another one) and/or coiling. The CC/SSC ratio which is about 50 μg/g in the upper half of the water column increases from nearly 0 to 50 μg/g in the lower half as function of time. This implies that a significant amount of algae is reaching the seafloor. The particles residing below the pycnocline are denser, as they are composed of flocs with a large residence time in the water column (and hence are more prone to be coiled) and more susceptible to contain larger amounts of mineral sediment, since more mineral sediment is to be found at the bottom of the water column. During HWS both algae-rich and mineral-rich flocs settle down but do not necessarily catch-up (limiting macroflocculation), which explains the large polydispersity of the measured PSD's. It has similarly been found, by the analysis of different European estuaries [31], that the relative ratio of size class 2a (usually called "microflocs") and size class 2b (called "macroflocs") do not depend on shear and that the system composed of Class 2a flocs and Class 2b flocs is a steady-state – which is another indication that macroflocculation in the water column is not a major process.

4. Conclusions

In this chapter, the dynamics of flocculation are discussed in connection to both laboratory and in-situ experiments and observations. Three classes of particles, defined by mass, size and settling velocity have been presented and are summarized in **Table 2**.

The equation to be implemented in a sediment transport model relates to the process of "microflocculation" whereby mineral sediment of Class 1 is aggregating with organic matter, creating a Class 2a or Class 2b floc. The rate of mass transfer between Class 1 and Class 2 can be obtained from laboratory experiments in closed vessels (to ensure mineral mass conservation during the experiment), and linked with changes in particle sizes over time. Studying flocculation in closed vessels is at

Type	Mineral sediment (unflocculated)	Mineral sediment flocculated with organic matter	
Class	1	2a	2b
Size	< 20 μm	[20–200] μm	>200 μm
Mass transfer between classes	$m_1(t)$ (mass mineral sediment free in suspension)	$m_2(t)$ (mass mineral sediment inside flocs) $dm_2/dt = -dm_1/dt$	
Density and settling velocity	2.6 kg/L [0–0.5] mm/s	[2.6 – 1.16] kg/L [0.5–10] mm/s	[1.16 – 1.02] kg/L [0.5–10] mm/s

Table 2.
Definitions of the classes with associated size, mass concentration, density and settling velocity.

present done in conditions that differs from in-situ conditions. The shear stresses in particular are usually higher in laboratory experiments (to avoid settling of particles in pipes or jars), the mineral clay concentrations are higher than in-situ (to ensure a proper detection by laser diffraction) and differential settling/flocculation of flocs in high water columns, with finite residence time, ought to be better studied. More work is also required to link the settling velocities obtained from settling column experiments via Eq. (13) to estimated settling velocities from in-situ techniques using Eq. (18). The one example given in the chapter shows that the velocities were different even though of the same order of magnitude.

The process of "macroflocculation" whereby a Class 2a or Class 2b is aggregating with another Class 2a or Class 2b is found to be a minor process in the water column (but might play a significant role close to the bed, where flocs interact more). As microflocculation is fast it is expected that at the top of the water column, where large particles of organic matter (like algae) are advected in summer, Class 2b particles are predominantly formed. Class 2a particles can be formed in regions where organic matter is less abundant, or where the shear is high, as by shearing flocs become denser and get a more spherical shape (by coiling). Class 2b particles can thus become Class 2a particles over time, as was demonstrated by laboratory experiments, and visible from under-water video microscopy. Another source of Class 2a particles is originating from resuspension from the bed, as it was observed that upon resuspension flocs are denser and smaller than before deposition. More work is required to parametrize the boundary condition (deposition/erosion) at the bed and in particular the mass transfer between Class 2 and Class 1 in the bed. This boundary condition is of course crucial for any sediment transport model.

Acknowledgements

This work has been performed in the frame of the grant NWO 869.15.011 "Flocs and Fluff in the Delta" (The Netherlands), the project 'Coping with deltas in transition' within the Programme of Strategic Scientific Alliance between China and The Netherlands (PSA), the Ministry of Science and Technology of People's Republic of China (2016YFE0133700), the Natural Science Foundation of China (51739005·U2040216), and the MUDNET academic network (https://www.tudelft.nl/mudnet/).

Author details

Claire Chassagne[1*], Zeinab Safar[1], Zhirui Deng[1,2], Qing He[2]
and Andrew J. Manning[1,3,4,5]

1 Faculty of Civil Engineering and Geosciences, Department of Hydraulic
Engineering, Delft University of Technology, The Netherlands

2 State Key Laboratory of Estuarine and Coastal Research, East China Normal
University, Shanghai, People's Republic of China

3 Coasts and Oceans Group, HR Wallingford, Oxon, UK

4 Energy and Environment Institute, University of Hull, Hull, East Riding of
Yorkshire, UK

5 School of Biological and Marine Sciences, University of Plymouth, Plymouth,
Devon, UK

*Address all correspondence to: c.chassagne@tudelft.nl

IntechOpen

References

[1] Hesse, Roland F., Anna Zorndt, and Peter Fröhle. "Modelling dynamics of the estuarine turbidity maximum and local net deposition." Ocean Dynamics 69.4 (2019): 489-507.

[2] Grasso, Florent, et al. "Suspended sediment dynamics in the macrotidal Seine Estuary (France): 1. Numerical modeling of turbidity maximum dynamics." Journal of Geophysical Research: Oceans 123.1 (2018): 558-577.

[3] Le Normant, C. "Three-dimensional modelling of cohesive sediment transport in the Loire estuary." Hydrological processes 14.13 (2000): 2231-2243

[4] Blumberg, A.F., Z-G Ji, and C.K. Ziegler. 1996. Modeling outfall plume behavior using far field circulation model. Journal of Hydraulic Engineering. ASCE, Vol. 122, No. 11

[5] Van Maren, D. S., et al. "Formation of the Zeebrugge coastal turbidity maximum: The role of uncertainty in near-bed exchange processes." Marine Geology 425 (2020): 106186.

[6] Shen, X., Lee, B. J., Fettweis, M., & Toorman, E. A. (2018). A tri-modal flocculation model coupled with TELEMAC for estuarine muds both in the laboratory and in the field. Water research, 145, 473-486.

[7] Maggi, F. (2009). Biological flocculation of suspended particles in nutrient-rich aqueous ecosystems. Journal of Hydrology, 376(1-2), 116-125.

[8] Lai, H., Fang, H., Huang, L., He, G., & Reible, D. (2018). A review on sediment bioflocculation: Dynamics, influencing factors and modeling. Science of the total environment, 642, 1184-1200.

[9] Mietta, F., Chassagne, C., Verney, R., & Winterwerp, J. C. (2011). On the behavior of mud floc size distribution:

model calibration and model behavior. Ocean Dynamics, 61(2-3), 257-271.

[10] Markussen, Thor Nygaard, and Thorbjørn Joest Andersen. "A simple method for calculating in situ floc settling velocities based on effective density functions." Marine Geology 344 (2013): 10-18.

[11] Smith, S. Jarrell, and Carl T. Friedrichs. "Image processing methods for in situ estimation of cohesive sediment floc size, settling velocity, and density." Limnology and Oceanography: Methods 13.5 (2015): 250-264.

[12] Manning, A. J., and K. R. Dyer. "A comparison of floc properties observed during neap and spring tidal conditions." Proceedings in Marine Science. Vol. 5. Elsevier, 2002. 233-250.

[13] Manning, Andrew J., Sarah J. Bass, and Keith R. Dyer. "Floc properties in the turbidity maximum of a mesotidal estuary during neap and spring tidal conditions." Marine Geology 235.1-4 (2006): 193-211.

[14] Manning, A.J., Whitehouse, R.J.S. and Uncles, R.J. (2017). Suspended particulate matter: the measurements of flocs. In: R.J. Uncles and S. Mitchell (Eds), ECSA practical handbooks on survey and analysis methods: Estuarine and coastal hydrography and sedimentology, Chapter 8, pp. 211-260, Pub. Cambridge University Press.

[15] Many, G., Bourrin, F., de Madron, X. D., Pairaud, I., Gangloff, A., Doxaran, D., ... & Jacquet, M. (2016). Particle assemblage characterization in the Rhone River ROFI. Journal of Marine Systems, 157, 39-51.

[16] Fettweis, Michael. "Uncertainty of excess density and settling velocity of mud flocs derived from in situ measurements." Estuarine,

Coastal and Shelf Science 78.2 (2008): 426-436.

[17] Russel, W. B., Russel, W. B., Saville, D. A., & Schowalter, W. R. (1991). Colloidal dispersions. Cambridge university press.

[18] Elimelech, M., Gregory, J., & Jia, X. (2013). Particle deposition and aggregation: measurement, modelling and simulation. Butterworth-Heinemann.

[19] Chassagne, C. (2020). Introduction to Colloid Science, Delft Academic Press, ISBN 9789065624376

[20] Chassagne, Claire, and Zeinab Safar. "Modelling flocculation: Towards an integration in large-scale sediment transport models." Marine Geology 430 (2020): 106361.

[21] Bubakova, P., Pivokonsky, M., and Filip, P. (2013). Effect of shear rate on aggregate size and structure in the process of aggregation and at steady state. Powder Technology, 235:540-549.

[22] P. Jarvis, B. Jefferson, J.Gregory, and S.A. Parsons. A review of floc strength and breakage. Water Research, 39:3121–3137, 2005.

[23] Manning, A. J., and K. R. Dyer. "A comparison of floc properties observed during neap and spring tidal conditions." Proceedings in Marine Science. Vol. 5. Elsevier, 2002. 233-250.

[24] Cahill, J., Cummins, P. G., Staples, E. J., & Thompson, L. (1987). Size distribution of aggregates in flocculating dispersions. Journal of colloid and interface science, 117(2), 406-414.

[25] Mietta, F. "Evolution of the floc size distribution of cohesive sediments", PhD thesis, Delft University (2010), ISBN 9789088911583

[26] Shakeel, Ahmad, Zeinab Safar, Maria Ibanez, Leon van Paassen, and Claire Chassagne. "Flocculation of clay suspensions by anionic and cationic polyelectrolytes: A systematic analysis." Minerals 10, no. 11 (2020): 999.

[27] Ibanez Sanz, M. E. Flocculation and consolidation of cohesive sediments under the influence of coagulant and flocculant. Diss. Delft University of Technology, 2018

[28] Manning, A. J., and K. R. Dyer. "Mass settling flux of fine sediments in Northern European estuaries: measurements and predictions." Marine Geology 245.1-4 (2007): 107-122.

[29] Dyer, K. R., and A. J. Manning. "Observation of the size, settling velocity and effective density of flocs, and their fractal dimensions." Journal of sea research 41.1-2 (1999): 87-95.

[30] Shen, X., Toorman, E.A., Lee, B.J., Fettweis, M., 2018b. Biophysical flocculation of suspended particulate matters in Belgian coastal zones. Journal of Hydrology 567, 238–252. https://doi.org/10.1016/j.jhydrol.2018.10.02

[31] Soulsby, R.L., Manning, A.J., Spearman, J., Whitehouse, R.J.S., 2013. Settling velocity and mass settling flux of flocculated estuarine sediments. Marine Geology 339, 1–12. https://doi.org/10.1016/j.margeo.2013.04.006

[32] Safar, Z. Suspended Particulate Matter formation and accumulation in the delta Diss. Delft University of Technology, 2021

[33] Maggi, F., Tang, F.H.M., 2015. Analysis of the effect of organic matter content on the architecture and sinking of sediment aggregates. Marine Geology 363, 102–111. https://doi.org/10.1016/j.margeo.2015.01.017

[34] Takabayashi, M., Lew, K., Johnson, A., Marchi, A., Dugdale, R., Wilkerson, F.P., 2006. The effect of nutrient availability and temperature on chain

length of the diatom, *Skeletonema costatum*. Journal of Plankton Research 28, 831–840. https://doi.org/10.1093/plankt/fbl018

[35] Deng, Zhirui, et al. "The role of algae in fine sediment flocculation: In-situ and laboratory measurements." Marine Geology 413 (2019): 71-84.

[36] Deng, Z., He, Q., Chassagne, C., & Wang, Z. B. (2021). Seasonal variation of floc population influenced by the presence of algae in the Changjiang (Yangtze River) Estuary. Marine Geology, 440, 106600

Chapter 6

Advances in Maintenance of Ports and Waterways: Water Injection Dredging

Alex Kirichek, Katherine Cronin, Lynyrd de Wit and Thijs van Kessel

Abstract

The main objective of this chapter is to demonstrate developments in port maintenance techniques that have been intensively tested in major European ports. As regular port maintenance is highly expensive, port authorities are considering alternative strategies. Water Injection Dredging (WID) can be one of the most efficient alternatives. Using this dredging method, density currents near the bed are created by fluidizing fine-grained sediments. The fluidized sediment can leave the port channels and be transported away from the waterways via the natural force of gravity. WID actions can be successfully coupled with the tidal cycle for extra effectiveness. In addition, WID is combined with another strategy to reduce maintenance dredging: the nautical bottom approach, which enables the vessel to navigate through the WID-induced fluid mud layer. The nautical bottom approach uses the density or the yield stress of sediment to indicate the navigability after WID rather than the absolute depth to the sediment bed. Testing WID-based port maintenance requires thorough preparation. Over the years modeling and monitoring tools have been developed in order to test and optimize WID operations. In this chapter, the application of the recently developed tools is discussed.

Keywords: fluid mud, dredging, sailing through mud, WID, nautical depth, cohesive sediment

1. Introduction

Navigation in ports, canals and waterways must be safeguarded by maintenance dredging to remove sediments deposited by tide, river flows and currents. In order to keep ports and waterways accessible, this non-contaminated sediment is typically dredged by a trailing suction hopper dredger (TSHD) and reallocated at sea [1].

Maintenance dredging of sediment deposits can be highly expensive and inefficient as it must be done on a regular basis. Therefore, port authorities seek tailor-made solutions to reduce the costs and at the same time guarantee safe navigation in ports and waterways. Over the last decades, a number of strategies for port maintenance have been tested by port and governmental authorities. Maintenance dredging can be optimized by techniques to avoid or reduce sedimentation, such as optimization of port design, current deflecting walls, see [2], or by designing a sedimentation trap to focus sediment deposition in order to make reallocation easier and to reduce sediment deposition in other port areas [3].

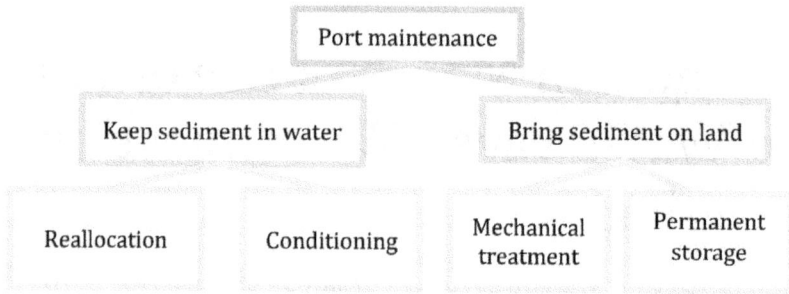

Figure 1.
Types of port maintenance methods which are based on the dredging methods keeping sediment in water or bringing dredged material on land.

Once dredging has conducted, typical strategies for dredged sediment management are either based on the concepts of keeping sediment in the water system or bringing sediment on land (see **Figure 1**). The former is generally considered as the most cost-effective strategy. However, the latter can be utilized for beneficial re-use of dredged sediment, thus better embedded into a circular economy.

It is a well-known fact that in major sea ports fine-grained sediment deposits are routinely reallocated from the port area either further away downstream from the dredged area or directly to the sea depending on the return flow of from the real-location locations. The choice in reallocation area often consists of finding a balance between minimizing sediment return flows back into the harbor and transport distance and costs. Often, the reallocation of dredged sediment is combined with sediment management within a building with nature concept [4]. These reallocation projects are mainly focused on the reallocation of fine-grained sediment for land creation or improvement, wild habitat restoration, shore nourishment and marsh or wetland development [5–7].

In contrast to reallocation of sediment, conditioning is used for port maintenance with the assumption that the sediment stays in the port area. The goal of conditioning the sediment is to create navigable conditions in waterways while keeping the sediment in place. In this case, the nautical bottom concept is often applied for navigation through mud [8–10]. One of the examples for applying sediment conditioning for port maintenance is in the Port of Emden. The sediment first dredged and then conditioned by reducing the strength of dredged sediment in the dredging vessel [8]. The created fluid mud is then pumped back to the port mouth creating a weak navigable fluid mud layer. If the transport of fluid mud towards the river equals the import of suspended mud by exchange flows, a dynamic equilibrium is achieved without residual import, hence dredging.

These techniques do not apply to contaminated dredged sediment which is either stored in confined disposal facilities [1, 11] or processed in sediment treatment facilities [12, 13]. The latter technology uses mechanical treatment to prepare the sediment for further beneficial re-use options. Recently, mechanical treatment is also used for non-contaminated sediment as dredged sediment is being recognized as a resource. The treated material can be used as a constructional component for building and re-enforcement of infrastructure [14, 15].

Water injection dredging (WID) can be used as a tool for both reallocation and conditioning of the deposited sediment. The efficiency of this dredging method has been recognized over the past 30 years. However, the successful application of WID can be only achieved by combining technical approaches with knowledge of the system where WID is to be applied. Particularly, the following key questions have to be answered in order to understand better the impact of WID on reallocation and conditioning of cohesive sediment:

- What type of sediment is to be relocated or conditioned by WID?

- What are the hydrodynamic conditions and bathymetry in the WID area?

- How fluidized sediment is distributed in port basins after WID?

- How far and where is the WID-induced plume transported after WID?

- What is the impact of WID on near-surface turbidity and how is this influenced by operational parameters?

- What criteria for navigation can be used in WID-conditioned areas?

The goal of this chapter is to provide an overview of the developed knowledge and tools that can be used for addressing the abovementioned questions. In addition, recently-developed numerical modeling, field and laboratory experiments can provide the necessary information for optimizing WID and defining the boundary conditions for its application. Finally, the recent findings on navigable conditions in ports and waterways, where WID is used for conditioning the sediment and keeping fluid mud in place, are discussed.

2. Working principles behind water injection dredging (WID)

2.1 Fluidization of fine-grained sediment

The principle of the water injection process is based on fluidization of deposited sediment by a water jet (see **Figure 2**). Water injection is performed by injecting large volumes of water (approx. 12,000 m³/h) under relatively low pressure (approx. 1-1.5 bar) from water jet nozzles, that are distributed over an equal distance on the jet [16, 17]. The injected water penetrates the cavities between the individual sediment particles weakening the forces between them and destroying the formed structure of the bed. The water-sediment mixture forms a fluid mud layer of about 0.5-3 m thickness right above the bed. Most investigations show that the sediment

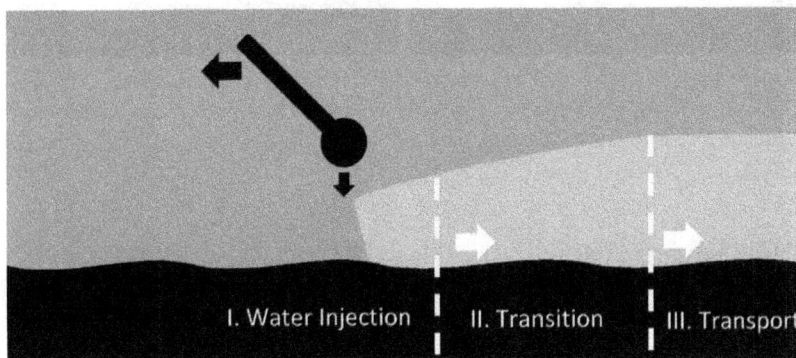

I. Water Injection | II. Transition | III. Transport

Figure 2.
Phases of WID: I. water injection and fluidization; II. Transition zone, where a density flow is created; III. Transport of the density flow. Adapted from [21, 22].

material hardly mixes into the upper water volume, and sediment transport of the fluidized mud layer remained predominantly close to the bottom [18, 19].

2.2 Transport of fluidized sediment

A sketch of WID performed in a navigational channel with a bed mainly consisting of fine-grained cohesive sediment is shown in **Figure 3**. The near bed fluidized sediment deposit generates a gravity driven density flow up to few meters high, transporting the sediment in a horizontal direction as a result of the density difference [17, 20–22]. This density flow can be described as a homogeneous suspension layer with a solid concentration of up to 200 g/l. Since the density between the fluid mud layer and the surrounding water body is different, fluid mud sets in motion under the action of natural hydrodynamic processes. Thus, WID is different from agitation dredging in which sediment is deliberately mixed over the full water column and then transported in horizontal direction as a passive plume by the ambient currents resulting in a less environmental-friendly outcomes.

The velocity of fluidized sediment is reported in the range between 0.3 m/s and 1 m/s [16, 21, 22]. Based on the hydrodynamic conditions in a port basin, WID-conditioned sediment can either settle over time in a low-energy area or be transported by means of gravity currents to deeper areas such as sediment traps [3].

Different transport distances from a few hundred meters to a few kilometers are reported for fluidized sediment [19, 21–23]. Natural transport of coarse-grained sandy sediment is substantially shorter. Therefore, the sediment composition of the bottom can be altered by WID operations. Fine-grained sediment can be generally more easily fluidized than coarse-grained material and has better transport properties. Since the fine grain fraction is transported away sooner and further than the coarse grain fraction, over time the particle size distribution of the sediment bed can be segregated as a result of dredging. Therefore, the coarse-grained component increases as a result of WID operations.

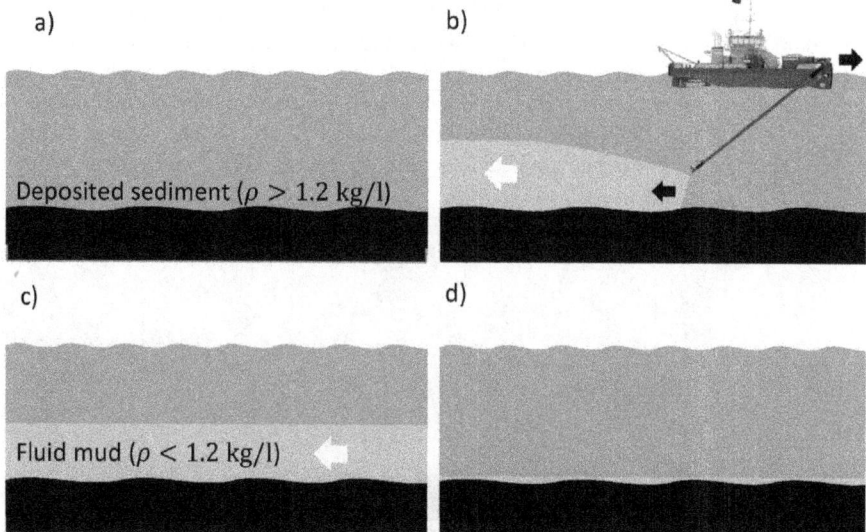

Figure 3.
Illustration of WID performed in a navigational channel during the ebb tide. a) Initial conditions for WID. b) Fluidization of deposited sediment during WID. c) WID-induced fluid mud layer. d) Final result after WID in case WID is conducted for sediment reallocation purposes.

2.3 Efficiency of WID

The effectiveness of the WID process can be influenced by various factors. The direction, velocity and achieved transport distances of the fluidized layer depend on the interaction of different physical forces. The important influencing factors are sediment composition and characteristics, WID operation characteristics, resulting density of the fluidized layer, bathymetry and natural currents and bed shear stresses in the WID area. The efficiency of the process is also influenced by the bathymetry of the dredging area and the prevailing natural currents. Productivity is generally increased when WID can be carried out so that fluid mud can flow with a natural gradient from higher to lower-lying bathymetry.

The composition and strength of the sediment are also essential for fluidization process. Although it is reported that WID has been also performed for removing coarse-grained sandy sediment and even consolidated soils [16, 21, 22], the best efficiency of WID has been achieved by fluidizing fine-grained sediment deposits. In [20] WID productions are reported in the order of a few thousand m^3/h for very fine-grained sediments and in the order of a few hundred m^3/h for coarser sediments.

The operational parameters for execution of WID are playing an important role for WID. The determining factors are the nozzles diameter, the flow velocity of the water from the jet, jet penetration, the forward movement of the jet pipe, and the distance between the jet nozzle and the surface of the sediment [24]. A WID operator can find the optimal combination of the aforementioned factors to achieve the maximum production of loosened material. However, not only the mass flux of loosened material should be optimized, but also the initial density, layer height and velocity. A thin but dense layer with little initial momentum will hardly spread, whereas a thick, diluted layer with high velocity will quickly mix with ambient water, with negative consequences for turbidity and focus of sedimentation footprint.

WID is generally considered as a relatively low-cost process [3, 25]. As the fluidized sediment is transported in the form of a density flow on the bed and is not distributed throughout the entire water column, WID is also characterized by a high level of environmental compatibility competing to traditional port maintenance dredging [3, 18]. Recently, it was also shown that WID is more CO_2 efficient than the regular TSHD maintenance because WID requires less fuel consumption than TSHD. All these aspects suggest that WID can be more attractive tool for port maintenance.

3. Modeling of WID

In recent years, different tools have been developed for optimizing WID processes and better prediction of sediment plume movement during WID. Numerical modeling tools can be used for estimating sediment dynamics in ports and waterways after WID.

Mid-field modeling is often used for calculating the sediment footprint on the areas up to about 1 km away from WID. The obtained knowledge on sedimentation can help to better design WID operations including real bathymetry of a navigational channel. Existing and hypothetical infrastructure can be included in mid-field modeling allowing for testing of WID in combination with sediment transport steering management solutions such as sediment traps, sills and current-deflection walls.

Far-field modeling evaluates the impact of WID on the scale of the entire port or estuary area. This kind of modeling is used for estimating WID reallocation strategies of sediment from the port basins to the sea and for assessing return flows. Simulations can demonstrate the transport of the WID plume during different phases of the tide and the impact of river and sea conditions. Based on the obtained information, the authorities can decide if conducting WID for reallocation purposes is effective in the port.

3.1 Mid-field modeling of WID

Mid-field modeling is carried out by two distinct models: a Lagrangian 1DV model and a 3D CFD model (TUDflow3D). The Lagrangian 1DV model is a rapid assessment tool which can be used for rather uniform bathymetry and slowly varying flows while neglecting lateral spreading. When these assumptions are not valid the more sophisticated 3D CFD model TUDflow3D can be used which includes lateral spreading and simulates a WID density current in three dimensions. TUDflow3D needs much more simulation time as the Lagrangian 1DV model.

The Lagrangian 1DV approach allows us to follow the development of the fluidized layer flow along a user-defined trajectory using a moving frame of reference. The 1DV model determines the thickness and the density (or the sediment concentration) of the fluidized mud layer and correlates these properties to the hydrodynamics in the water column and the slope of the bed. Additionally, it determines the sedimentation flux on the bed. For an equal initial momentum of the fluidized mud layer, the layer will flow further along a downward slope than along a flat bed. In general, the results of 1DV modeling can be used for a better planning of WID.

Figures 4 and **5** illustrate an example of utility of the 1DV model for water injection dredging. In both figures, the left panel shows the distribution of the sediment concentration and the height of the fluidized mud layer along the slope. The right panel shows the flow velocity of the fluidized mud layer. **Figure 4** shows the simulation of WID for an initial WID plume height of 2 m and **Figure 5** shows the results of WID for an initial WID plume height of 3 m. Both cases start with an initial sediment concentration of 170 kg/m^3 and 0.7 m/s flow velocity. It can be seen that a higher fluidized mud layer travels faster and reaches a higher internal velocity.

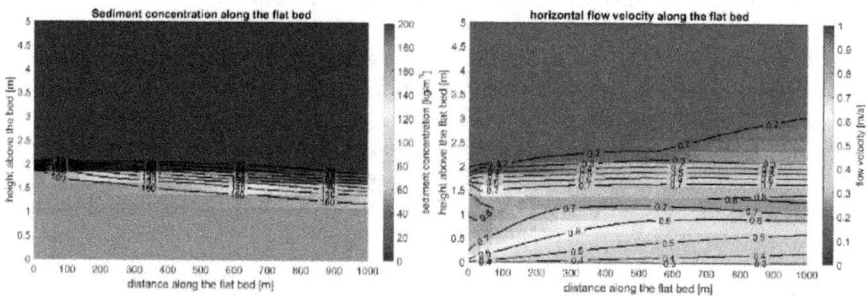

Figure 4.
1DV results for initial WID plume height of 2 m.

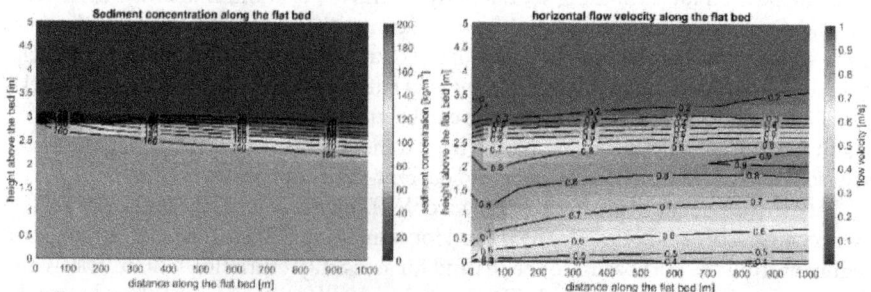

Figure 5.
1DV result for initial WID plume height of 3 m.

WID density-driven plumes can be also simulated in 3D by the CFD model TUDflow3D [26, 27]. Originally, TUDflow3D has been developed for accurate near field simulations of Trailing Suction Hopper Dredger overflow plumes on real scale. It has also been used for MFE (Mass Flow Excavation) plumes, deep sea mining tailing plumes and salinity driven density flows. TUDflow3D can supplement the 1DV model for complex situations in which the simplifications of the 1DV model make application impossible. TUDflow3D is fully 3D with variable density taken into account in all three dimensions (not just in the vertical), non-hydrostatic pressure and turbulence captured by either the accurate LES (Large Eddy Simulation) approach or by a faster RANS (Reynolds Averaged Navier Stokes) approach.

An instantaneous snapshot of the modeled density current is shown in **Figure 6**. The individual turbulent eddies and whirls resolved on the grid in LES are clearly visible. Comparison for time averaged velocity and Suspended Sediment Concentration (SSC) profiles with measured ones is given in **Figure 7**. Here, different manners of capturing turbulence are compared. In addition to LES with the WALE sub-grid-scale model, the RANS with Realizable K-Epsilon model and Realizable K-Epsilon model with reduced eddy viscosity near the bed are tested. In the latter the eddy viscosity near the bed is adjusted, effectively reduced, to correspond to the correct amount of bed shear stress. The results show that this adjustment improves the Realizable K-Epsilon results for this flow. The vertical SSC profile and layer thickness of the density current is captured very well in the CFD LES model and the velocity profiles are captured reasonably well with a small overprediction of the near bed velocity. The Realizable K-Epsilon results with adjusted near bed viscosity are considerably better as the default Realizable K-Epsilon results.

An example a of application of TUDflow3D for WID is given in **Figure 8**. In this CFD run a WID works along a 300 m track which it has done 6 times in a row.

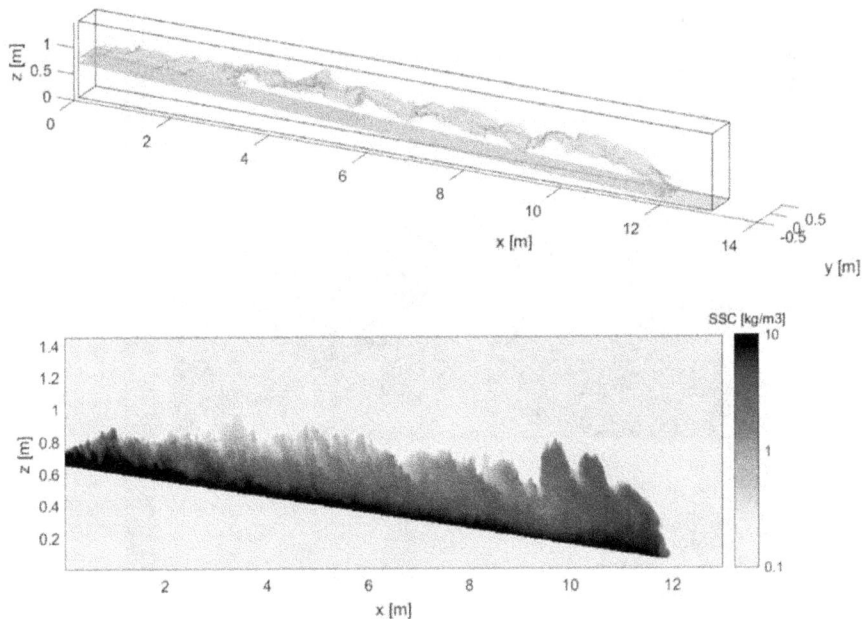

Figure 6.
Instantaneous LES snapshot of 3D contour (top) of a turbidity current and SSC at a vertical slice through the center of the turbidity current (bottom).

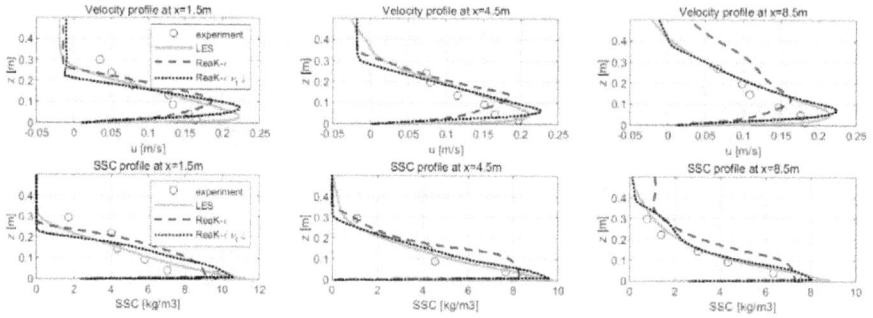

Figure 7.
Comparison modeled time averaged velocity and SSC profiles with 3 different turbulence settings (LES; realizable K-epsilon and realizable K-epsilon with reduced near bed viscosity) and measurements from [28].

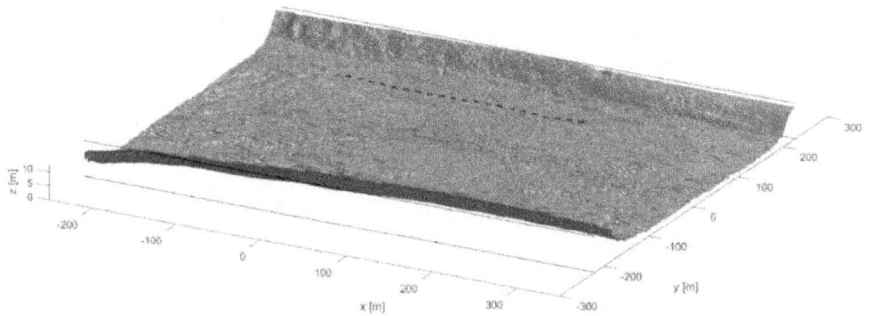

Figure 8.
Example of TUDflow3d simulation: Plume distribution from WID action along black dashed line.

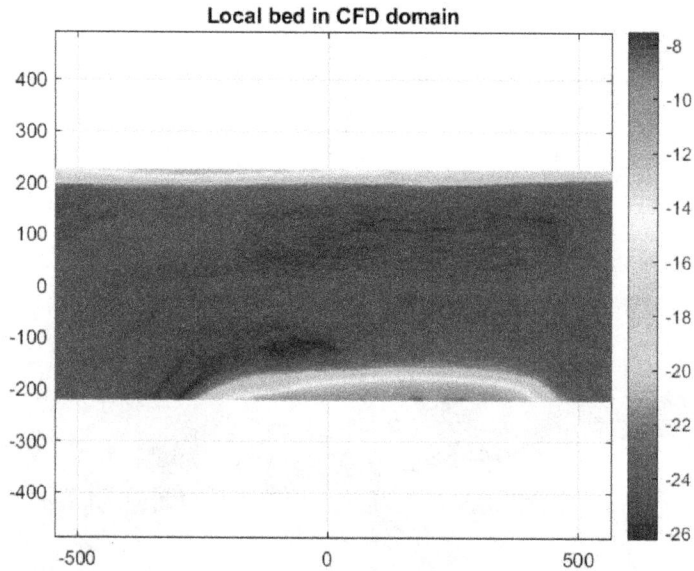

Figure 9.
Example of TUDflow3D simulation: Implementing bathymetry in a CFD domain.

The CFD model uses the real bathymetry of the port. The resulting WID plume is shown in brown and the bathymetry is illustrated as a gray surface. At the moment of this image the WID has just finished the 6th time along the black dashed track of 300 m long. In this example the WID plume flows down the sloping bed in lateral direction under influence of gravity. A top view of the bathymetry is shown in **Figure 9**.

A comparison of TUDflow3D and the Lagrangian 1DV model for WID in a lateral confined situation without bed-slope is shown in **Figure 10**. For this simulation, the following initial conditions were applied: initial WID layer thickens of 2 m, 170 kg/m³ and 0.7 m/s inflow (resulting in an influx of 238 kg/s). The example shows the simulated vertical velocity profiles and density profiles at different distances from the WID. The model also calculates the sedimentation flux out of the WID density current. The results of the 1DV model and full 3D CFD are close to each other for this case. For cases where the assumptions of the Lagrangian 1DV model (neglecting lateral spreading and slowly varying flow conditions) hold it is much faster as the more sophisticated TUDflow3D model and in other cases it is advised to use a 3D near field model like TUDflow3D.

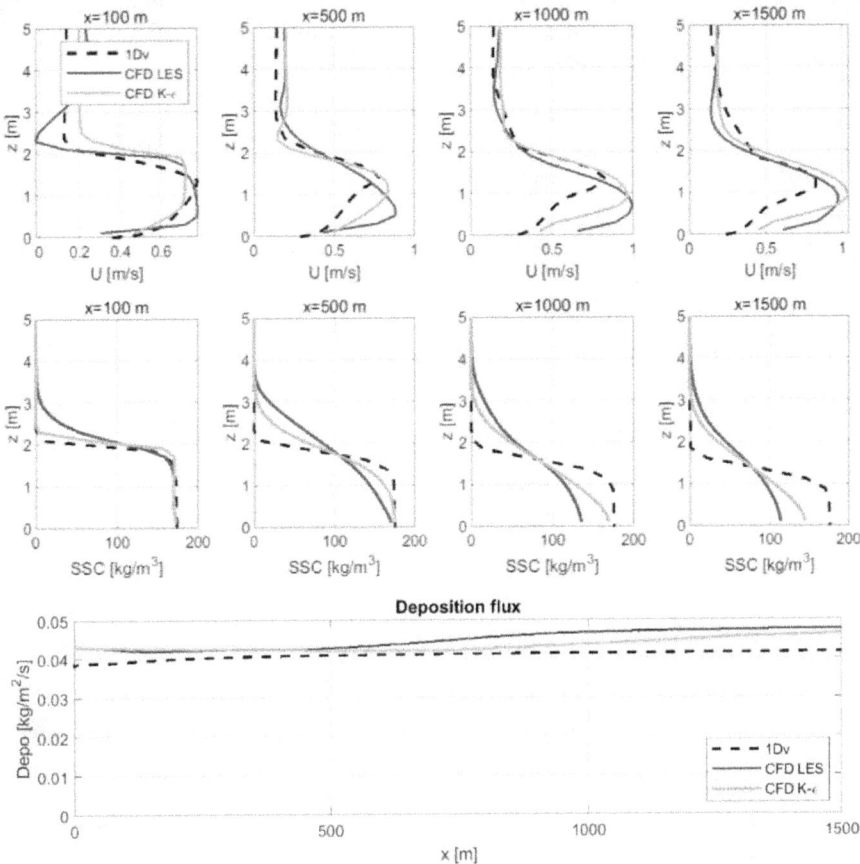

Figure 10.
Comparison of CFD model TUDflow3D and 1Dv simulations for WID in a lateral confined situation. TUDflow3D is compared for two different turbulence settings (LES; realizable K-epsilon).

3.2 Far-field modeling of WID

Sediment dynamics and specifically, the siltation of mud, in ports is of great interest to those responsible for the maintenance of ports, harbors and access channels around the world. The amount of siltation determines the frequency and volume of maintenance dredging needed to maintain navigable depth. In order to understand sediment dynamics in the system, in particular the processes responsible for suspended mud and fluid mud transport, a range of spatial and temporal scales must be analyzed. A numerical model is an ideal tool with which to investigate both the transport, deposition, and potential resuspension of a WID plume. Such a model was developed, using Delft3D, for the Rhine Meuse Delta in the Netherlands, in order to calculate both background fine sediment dynamics in the Port of Rotterdam and the transportation of a fluid mud layer after a WID operation.

Deltares' open source software Delft3D is a flexible, integrated modeling framework which simulates two and three-dimensional flow, waves, sediment transport and morphology (as well as dredging and dumping) on a time-scale of days to decades. The sediment transport module includes both suspended and bed/total load transport processes for an arbitrary number of cohesive and non-cohesive sediment fractions. It can keep track of the bed composition to build up a stratigraphic record. The suspended load solver is connected to the 2D or 3D advection–diffusion solver of the hydrodynamic module and importantly for fluid-mud simulations, density feedback can also occur.

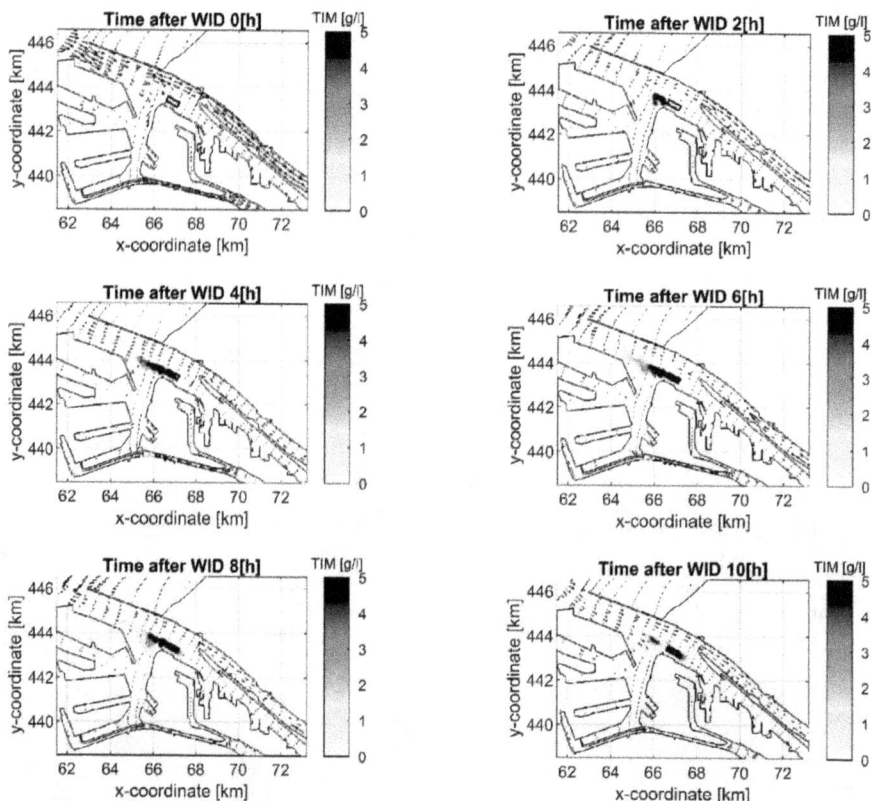

Figure 11.
Horizontal near bed plume spreading, WID starts 1 h before HW with a production rate of 500 kg/s.

For this work, a Delft3D model of the entire Rhine Meuse Estuary was setup. Hydrodynamic conditions were simulated for a full month, including wave effects. This hydrodynamic model is then used to force the sediment transport model. Background sediment concentrations are included in the model using three sediment fractions to represent the appropriate range of coarser and finer fractions. Once natural dynamics regarding sediment transport and sediment deposition in the different ports was captured, a range of WID tests could be undertaken. The parameters derived for different WID production rates in the mid-field modeling (described in Section 3.1) are used to define the initial conditions for the WID plume in the far-field model. Numerical experiments could then be performed such as simulating where the WID plume is transported to, the amount of return flow into different parts of the port and the amount of mixing that occurs throughout the water column. Vertical mixing may result in elevated turbidity levels near the surface, which should remain within the environmental limits. The model is also used to investigate the optimum location for sediment traps to capture the WID high density plume.

Figures 11 and **12** show an example of how the far-field modeling was used to investigate the impact of carrying out WID at different stages of the tidal cycle. WID was carried out in the area of a black rectangle. The colourbar indicates the distribution of suspended sediment concentration (SSC) in the port area. The duration of WID was 8 hours with a production rate of 500 kg/s. During 2 simulations, WID was initiated 1 h before high water (HW) and 1 h before low water (LW). The results of both simulations are shown in **Figures 11** and **12**, respectively.

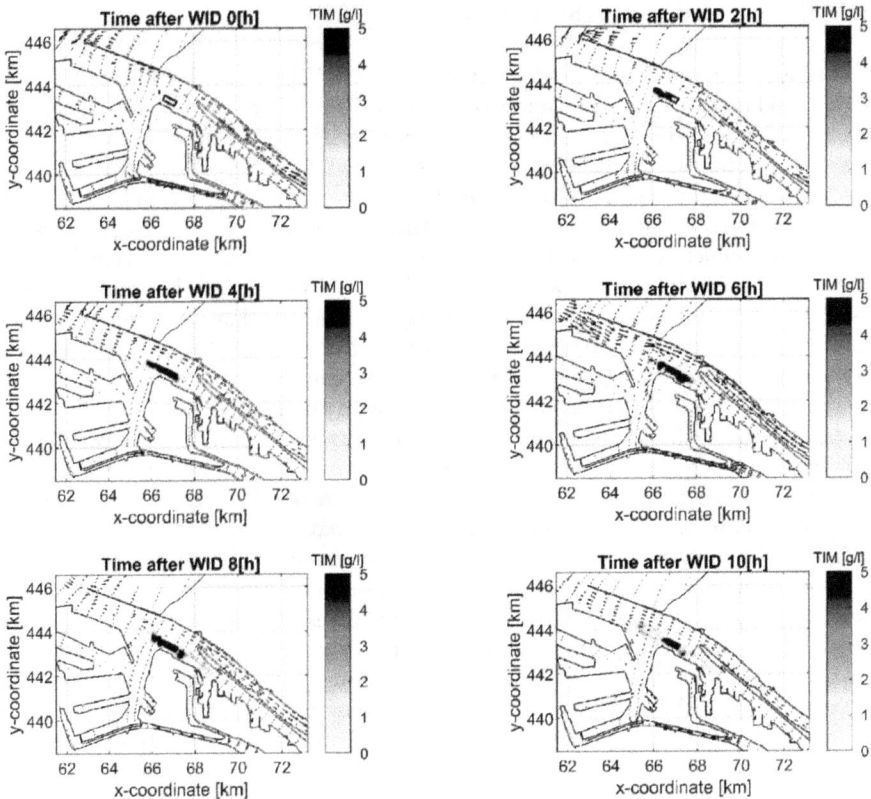

Figure 12.
Horizontal near bed plume spreading, WID starts 1 h before LW with a production rate of 500 kg/s.

The plume disperses in a distinct way between the simulation starting before high water (HW) compared to a start at low water (LW). **Figure 11** shows that the plume is predominantly dispersed in the seaward direction with the outgoing tide. For WID, this would be the most preferable conditions because in this way the suspended sediment will be relocated from the area where WID is conducted off-shore. However, after approximately six hours the flow is reversed, and the plume is pushed in the landward direction.

Figure 12 show the initial plume dispersion for the simulations in which sediment is released just before LW. The dispersion of the plume in the first 2 hours of the simulations is similar to the experiment with WID release just before HW. However, between four and eight hours a predominant landward plume dispersion is observed. After the flow reversal, it is observed that the plume starts to disperse in the seaward direction. A continuation of the landward spreading is observed in the channel because of the predominant landward flood directed current.

The far-field modeling illustrates the importance of the hydrodynamic conditions during WID. This knowledge can help to choose the most-efficient strategy for WID in ports and waterways with mud layers. The most efficient strategy is not only related to optimizing the sedimentation footprint, but also to minimizing vertical mixing and the contribution of WID to turbidity higher up in the water column. By choosing operational parameters wisely and executing WID operations only during favorable hydrodynamic conditions demands on sedimentation footprint and turbidity are more easily met.

4. WID and navigation through mud

In low-energy regions or in a tidal area of the port, WID-induced sediment can form a fluid mud layer that remains in the port area. The thickness of WID-induced fluid mud layer is often larger than the thickness of original mud layer resulting in a reduced draft for the incoming vessels. In this case, WID is often combined with the nautical bottom approach defined by PIANC for navigation. According to PIANC, 'The nautical bottom is the level where physical characteristics of the bottom reach a critical limit beyond which contact with a ship's keel causes either damage or unacceptable effects on controllability and manoeuvrability' [10, 29]. The nautical bottom allows to use the fluid mud in estimates of under keel clearance (UKC) that the vessels can navigate in the port areas with no unacceptable effects on controllability and maneuvering of the vessels. If accepted by the port authorities, the nautical bottom approach is used for navigation through mud in ports and waterways with fluid mud layers.

Generally, the density of the top sediment layer is used for defining the nautical bottom (see **Figure 13**). The level, where the density of sediment is lower than 1.2 t/m^3, is widely accepted for navigation in ports. Ports in Rotterdam, Zeebrugge, Bordeaux, Saint-Nazaire, Bristol, Bangkok, Tianjin have successfully adapted the density criterium for navigation [29, 30]. However, the Port of Emden relies on the rheological properties rather than density of the sediment for defining the nautical bottom. The yield stress of the top sediment layer gives an indication if the sediment is navigable or not. The sediment with yield stress lower than 100 Pa is considered navigable. The choice of the nautical bottom criterium is related to the conditioning of sediment, that the Port of Emden has been conducting for port maintenance.

The knowledge on in-situ density or rheological properties of the top sediment layer are necessary for implementing the nautical bottom approach. There are in-situ tools that can provide an information about vertical profiles of density and strength in water-mud column. The in-situ devices Rheotune, Graviprobe and DensX have been intensively tested for the nautical bottom approach over last years [3, 29, 31].

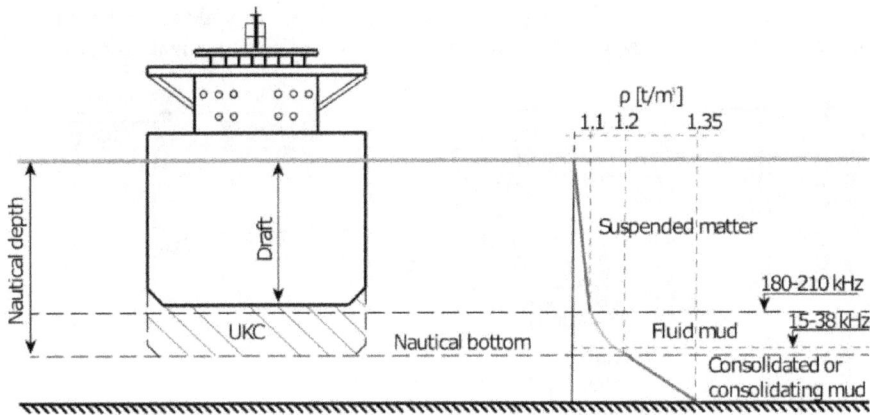

Figure 13.
Illustration of the nautical bottom concept with the density of 1.2 t/m³.

An example of in-situ measurement of density and yield stress provided by Rheotune is shown in **Figure 14**. The measurements are conducted in a sediment trap that was filled with WID-induced fluid mud during day 1. The development of density and yield stress of WID-induced sediment has been observed for the period of 3 months. The in-situ devices can naturally provide only 1D vertical profiles. However, the thickness of mud layer can be defined from the profiles if the critical value for physical parameters is defined.

In the example given in **Figure 15**, the critical value for the density is chosen as 1.2 t/m³ providing the density-based nautical bottom shown in red line. In this case, the SILAS software is used for matching the density given by Rheotune (shown by vertical blue line in **Figure 15**) to the seismic data of 38 kHz. The measurements are conducted 7, 21 and 42 days after WID.

The development of WID-induced mud layer be also estimated with the numerical code solving the Gibson Eq. [32]. For instance, settling and consolidation of fluid mud can be predicted by matching the measured data to the model output. **Figure 16** shows the comparison of 1DV model and measured data during consolidation of WID-induced fluid mud layer. The model's output is the density of mud and the water mud interface as a function of time, that can be correlated to measured densities and multibeam data, respectively. The latter can typically provide a reliable water-mud interface for WID operations. For instance, **Figure 17** shows the development of water-mud interface before, during and after WID in the Calandkanaal.

Vertical density profiles are shown in the right panel of **Figure 16**. The density measurements can be done by different penetrometers [3, 31, 33], in this case the densities are measured by DensX. It can be observed that the measured density profiles show a good resemblance with the results of numerical modeling [31, 33]. Thus, the combination of the model with the in-situ measurements can potentially be used for predicting the development of the nautical bottom in time.

An example of the application of PIANC's nautical bottom approach after WID in the Port of Rotterdam is shown in **Figure 18**. The standard multibeam echosounder survey indicated the bathymetry that corresponds to the water-mud level. However, the WID-induced fluid mud has relatively low densities (<1200 kg/m³) and weak strength (<100 Pa). Therefore, the nautical bottom approach can be applied. Adapting either a density-based (1200 kg/m³) or yield stress-based (100 Pa) criterium for the nautical bottom results in an additional 1.5 and 2 m of navigable depth, respectively.

20 days after WID, these differences are reduced. However, the yield stress-based nautical bottom still shows an advantage of about 0.5 m of extra navigable depth.

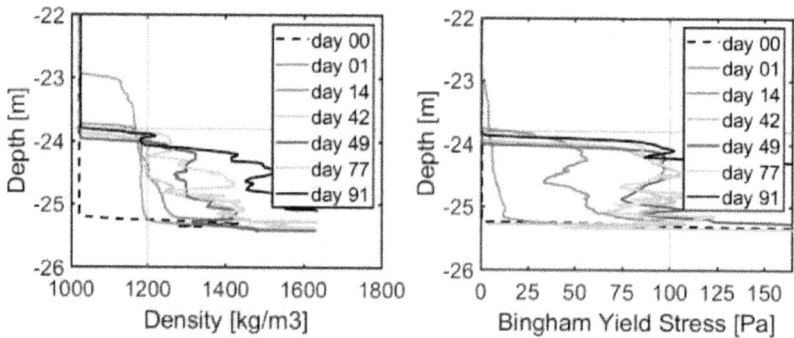

Figure 14.
Density and yield stress profiles measured by Rheotune.

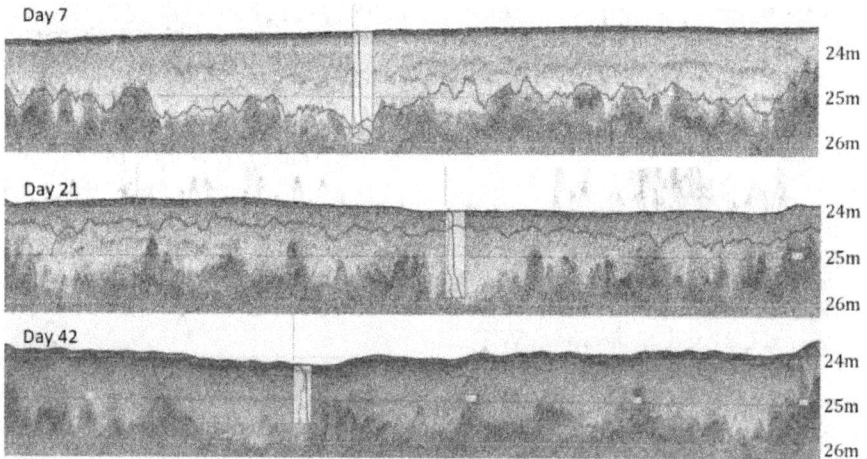

Figure 15.
Development of the density-based nautical bottom after WID. Red line shows the level, where the density of sediment is equal to 1.2 t/m³.

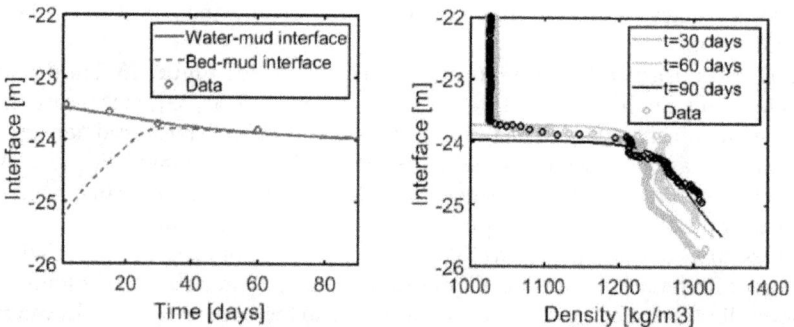

Figure 16.
Estimating consolidation of fluidized mud layer after WID. Left panel shows development of water - fluidized mud interface as well as fluidized mud – Consolidated bed interface. Right panel show model predictions (solid lines) and in-situ measurements (symbols) of densities in water-mud vertical column.

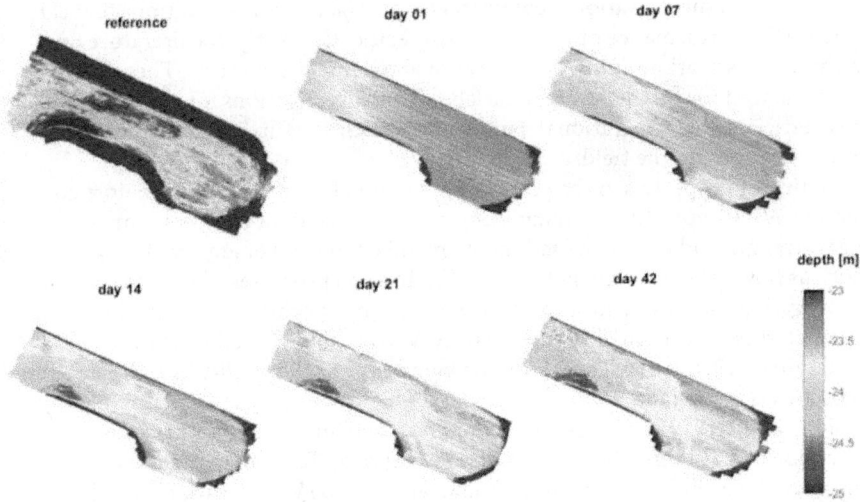

Figure 17.
Multibeam measurements indicating water-mud interface before WID (reference), during WID (day 1) and after WID (day 7 - day 42) in the Calandkanaal.

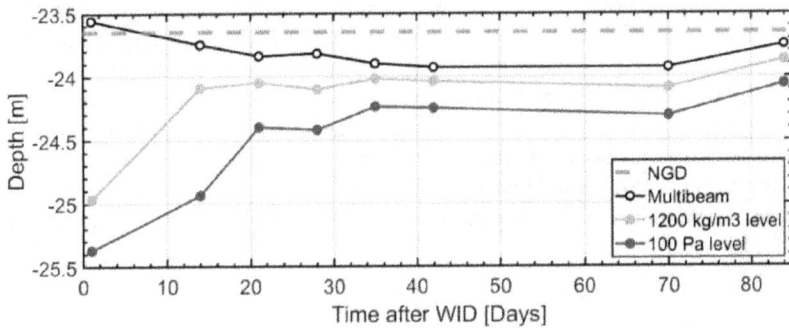

Figure 18.
An example of applying the nautical bottom approach after WID [3]. The density-based (1200 kg/m³) or yield stress-based (100 Pa) criteria brings additional 2 m for nautical depth comparing to the standard multibeam-based navigational criterium.

5. Discussion

Water injection dredging is a widely applicable dredging method. The efficiency of the method for maintaining ports and waterways is generally high. WID operational parameters, knowledge of sediment properties, boundary and hydrodynamic conditions of the maintained area can greatly increase the efficacy of the water injection process. The most important parameters and factors influencing the performance of WID are the following: WID operational parameters (diameter of nozzles, flow velocity from the nozzle, stand-off distance of the jet, trailing speed of the WID vessel), sediment properties (grain size distribution, shear strength, density, oxygen consumption potential and sediment quality), boundary conditions of the maintained area (bathymetry, slope angle, embankments), hydrodynamics conditions (direction and velocity of tidal currents, existing density currents and salinity gradients).

Apart from the operational parameters, other factors and conditions that can increase the performance of WID are site-specific. Currently, the literature on research investigations into WID operational parameters is scarce. Therefore, there is a need for further systematic laboratory investigations for exploring the most-efficient WID operational parameters, which can further maximize the WID production rates in the field.

Sediment properties in the proposed area for WID can be studied before conducting WID. Typically, sediment samples are collected for laboratory analysis. The shear strength and density of sediment are linked to WID operational parameters (such as flow velocity) during the WID fluidization processes. The literature on investigations of sediment properties while testing varying WID operational parameters is very limited. Predominantly, WID is applied in the area with no-contaminated sediment. Therefore, the knowledge of the quality of sediment in the WID area is important.

The geometry of the WID area should be taken into account for planning and execution of WID operations in port and waterways. Bathymetric charts, which, will provide the information about deeper areas in the WID location, which are typically filled in with fluid mud after WID. Furthermore, bathymetric charts will indicate the slopes in the WID area, which can be also used for transporting the fluidized mud more efficiently.

Hydrodynamic conditions in the WID area should be taken into account when determining the final fate of fluid mud generated by WID, whether WID is used for the transport or conditioning of mud. For the transport of mud, the knowledge of the direction of the natural current and current velocities can help to minimize the spread of the WID-induced fluid mud deeper into the port area and maximize the transport of the sediment from the port area. For the conditioning of mud, the hydrodynamic conditions can potentially provide an indication whether fluid mud starts to settle in the allocated area or is transported to other locations of the port. Salinity gradients and local density currents can influence the density currents by damping the velocity of WID-induced fluid mud, thus decreasing production rates in the WID-area.

6. Conclusions

This chapter focusses on presenting an overview of developed knowledge for WID. In particular, new insights gained using a combination of in-situ monitoring and numerical modeling. The research focusea on fluid mud behavior and transport, but also the resulting sediment plume. Both mechanisms are important and depend on the surrounding hydrodynamic conditions. Mid-field modeling was used to investigate the WID plume flow and deposition behavior up to 1 km away from the WID dredger. The WID-induced fluid mud layer thickness and WID production estimates were used as input in to the far-field model. The far-field model was used to determine where the WID-induced plume traveled under different tidal and discharge conditions, how much deposited back in the harbors and how much was flushed out to sea with the ebb tide. The model was also used to test different disposal locations to reduce return flow.

Key factors and parameters influencing the efficiency of WID have been identified from the available literature and discussed further. The modeling tools presented in the chapter can potentially help to analyze the sediment properties, boundary conditions and hydrodynamic conditions in the WID area and in the entire port area. However, more experimental research is needed for defining the most-efficient set of operational parameters. Particularly, the knowledge on linking WID operational parameters with sediment properties for maximizing production rates is very scarce.

By combining measurements from the field, laboratory experiments on fluid mud properties, with a state-of-the-art modeling approach, new insights were gained on the best approach for implementing WID as a maintenance dredging strategy. In addition due to more efficient maintenance, reduction of costs, CO_2 emissions and additional environmental impacts is achieved during the application of these techniques.

Acknowledgements

The work in this study is funded by the Port of Rotterdam and by Topconsortium voor Kennis en Innovatie (TKI) Deltatechnologie subsidy. The research is carried out within the framework of the MUDNET academic network https://www.tudelft.nl/mudnet/

Conflict of interest

The authors declare no conflict of interest.

Nomenclature

WID	water injection dredging
TSHD	trailing suction hopper dredger
RANS	Reynolds averaged Navier Stokes
SSC	suspended sediment concentration
LES	Large Eddy Simulation
MFE	Mass Flow Excavation
HW	high water
LW	low water
PIANC	World Association for Waterborne Transport Infrastructure
UKC	under keel clearance

Author details

Alex Kirichek[1,2*], Katherine Cronin[2], Lynyrd de Wit[2] and Thijs van Kessel[2]

1 Faculty of Civil Engineering and Geosciences, Delft University of Technology, The Netherlands

2 Deltares, The Netherlands

*Address all correspondence to: o.kirichek@tudelft.nl

IntechOpen

References

[1] Kirichek A, Rutgers R, Wensveen M, Van Hassent. Sediment management in the Port of Rotterdam. In: Proceedings of the 10th Rostocker Baggergutseminar; 11-12 September 2005; Rostock

[2] Kirby R. Minimising harbour siltation—findings of PIANC Working Group 43. Ocean Dynamics. 2011; 61: 233-244. https://doi.org/10.1007/s10236-010-0336-9

[3] Kirichek A, Rutgers R. Monitoring of settling and consolidation of mud after water injection dredging in the Calandkanaal. Terra et Aqua. 2020; 160:16-26

[4] Sittoni L, van Eekelen E, van der Goot F, Nieboer H. The living lab for mud: integrated sediment management based on building with nature concepts. In: Proceedings of the 22nd World Dredging Congress & Exposition (WODCON XXII); 2-26 April 2019; Shanghai

[5] SURICATES - Sediment Uses as Resources In Circular And Territorial EconomieS [Internet]. 2017. Available from: https://www.nweurope.eu/projects/project-search/suricates-sediment-uses-as-resources-in-circular-and-territorial-economies/ [Accessed: 2020-12-28]

[6] Baptist JM, Gerkema T, van Prooijen BC, van Maren DS, van Regteren M, Schulz K, Colosimo I, Vroom J, van Kessel T, Grasmeijer B, Willemsen P, Elschot K, de Groot AV, Cleveringa J, van Eekelen EMM, Schuurman F, de Lange HJ, van Puijenbroek MEB, Beneficial use of dredged sediment to enhance salt marsh development by applying a 'Mud Motor', Ecological Engineering. 2019; 127: 312-323.

[7] Marker Wadden - Natuurmonumenten. 2019. Available from: https://www.natuurmonumenten.nl/projecten/marker-wadden/english-version [Accessed: 2020-12-28]

[8] Wurpts R, Torn P. 15 Years' Experience with Fluid Mud: Definition of the Nautical Bottom with Rheological Parameters. Terra et Aqua. 2005. 99:22-32

[9] Kirichek A, Chassagne C, Winterwerp H, Vellinga T. How navigable are fluid mud layers? Terra et Aqua. 2018; 151:6-18

[10] PIANC. Harbour Approach Channels - Design Guidelines, Report 121, PIANC, Brussels, 2014.

[11] Slibdepot IJsseloog. 2010. Available from: https://web.archive.org/web/20100604043705/http://www.rijkswaterstaat.nl/water/plannen_en_projecten/vaarwegen/ketelmeer/ketelmeer/ketelmeer_oost/ [Accessed: 2020-12-28]

[12] METHA - MEchanical Treatment and Dewatering of HArbor sediments. 2020. Available from: https://www.hamburg-port-authority.de/en/themenseiten/metha/ [Accessed: 2020-12-28]

[13] AMORAS - Antwerpse Mechanische Ontwatering, Recyclage en Applicatie van Slib. 2020. Available from: https://www.maritiemetoegang.be/amoras [Accessed: 2020-12-28]

[14] Kleirijperij. 2019. Available from: https://eemsdollard2050.nl/project/pilot-kleirijperij/ [Accessed: 2020-12-28]

[15] Brede Groene Dijk. 2019. Available from: https://eemsdollard2050.nl/project/brede-groene-dijk/ [Accessed: 2020-12-28]

[16] Clausner JE. Water injection dredging demonstrations. U. S. Army Corps of Engineers, Waterways Experiment Station. Dredging Research. 1993. Vol. DRP-93-3

[17] Winterwerp JC, Wang ZB, van Kester JATM, Verweij JF. Far-field impact of water injection dredging in the Crouch River. Proceedings of the Institution of Civil Engineers Water & Maritime Engineering. 2002; 154 (4), 285-296.

[18] Borst WG, Pennekamp JGS, Goossens H, Mullié A, Verpalen P, Arts T, van Deumel PF, Rokosch WD. Monitoring of water injection dredging, dredging polluted sediment. In: Proceedings of the second international conference on dredging and dredged material placement (Dredging '94); 13-16 November; Walt Disney World, Lake Buena Vista, Florida. New York: ASCE; 1994. Vol. 2: p. 896-905

[19] Meyer-Nehls R, Gönnert G, Christiansen H, Rahlf H. Das Wasserinjektionsverfahren – Ergebnisse einer Literaturstudie sowie von Untersuchungen. In Hamburger Hafen und in der Unterelbe. 2000. ISSN 0177-1191

[20] PIANC - Injection Dredging, Report 120, PIANC, Brussels, 2013

[21] Verhagen HJ. Water injection dredging. In: Proceedings of the 2nd International Conference Port Development and Coastal Environment (PDCE 2000); 5-7 June 2000, Varna, Bulgaria, 2000.

[22] Murphy AM. DRP site visit: Water injection dredging. U. S. Army Engineer Waterways Experiment Station, Vicksburg, Miss. Dredging Research. 1993; DRP-93-1

[23] Nasner H. Injektionsbaggerung von Tideriffeln. Hansa. 1992; 129 (2): 195-196.

[24] Estourgie ALP. Theory and practice of water injection dredging. Terra et Aqua. 1988; 38:21-28

[25] Bray RN. Maintenance dredging: where do we go from here? The Dock & Harbour Autority LXX. 1989; 810:57-60

[26] de Wit L, Bliek A.J., van Rhee C. Can surface turbidity plume generation near a Trailing Suction Hopper Dredger be predicted? Terra et Aqua. 2020 September, 6-15

[27] de Wit L. 3D CFD modelling of overflow dredging plumes. [thesis]. Delft: Delft University of Technology; 2015. https://doi.org/10.4233/ uuid:ef743dff-6196-4c7b-8213-fd28684d3a58

[28] Parker G, Garcia M, Fukushima Y, Yu W. Experiments on turbidity currents over an erodible bed. J. of Hydraulic Research. 1987; 25(1):123-147. DOI: 10.1080/00221688709499292

[29] Kirichek A, Chassagne C, Winterwerp H, Vellinga T. How navigable are fluid mud layers? Terra et Aqua. 2018; 151: 6-18.

[30] McAnally WH, Teeter A, Schoellhamer D, Friedrichs C, Hamilton D, Hayter E, Shrestha P, Rodriguez H, Sheremet A, Kirby R. Management of Fluid Mud in Estuaries, Bays, and Lakes, Part 2: Measurement, Modeling, and Management. J. Hydraul. Eng. 2007. 133 (1).

[31] Kirichek A, Shakeel A, Chassagne C. Using in situ density and strength measurements for sediment maintenance in ports and waterways. J. Soils Sediments. 2020; 2546-2552. DOI: 10.1007/s11368-020-02581-8

[32] Merckelbach LM. Consolidation and strength evolution of soft mud layers. [thesis]. Delft: Delft University of Technology; 2000

[33] Kirichek A, Rutgers R. Water injection dredging and fluid mud trapping pilot in the Port of Rotterdam. In: Proceedings of the CEDA Dredging Days 2019, 7-8 November 2019; Rotterdam

Non-Intrusive Characterization and Monitoring of Fluid Mud: Laboratory Experiments with Seismic Techniques, Distributed Acoustic Sensing (DAS), and Distributed Temperature Sensing (DTS)

Deyan Draganov, Xu Ma, Menno Buisman, Tjeerd Kiers, Karel Heller and Alex Kirichek

Abstract

In ports and waterways, the bathymetry is regularly surveyed for updating navigation charts ensuring safe transport. In port areas with fluid-mud layers, most traditional surveying techniques are accurate but are intrusive and provide one-dimensional measurements limiting their application. Current non-intrusive surveying techniques are less accurate in detecting and monitoring muddy consolidated or sandy bed below fluid-mud layers. Furthermore, their application is restricted by surveying-vessels availability limiting temporary storm- or dredging-related bathymetrical changes capture. In this chapter, we first review existing non-intrusive techniques, with emphasis on sound techniques. Then, we give a short review of several seismic-exploration techniques applicable to non-intrusive fluid-mud characterization and monitoring with high spatial and temporal resolution. Based on the latter, we present recent advances in non-intrusive fluid-mud monitoring using ultrasonic transmission and reflection measurements. We show laboratory results for monitoring velocity changes of longitudinal and transverse waves propagating through fluid mud while it is consolidating. We correlate the velocity changes with shear-strength changes while the fluid mud is consolidating and show a positive correlation with the yield stress. We show ultrasonic laboratory results using reflection and transmission techniques for estimating the fluid-mud longitudinal- and transverse-wave velocities. For water/mud interface detection, we also use distributed acoustic sensing (DAS) and distributed temperature sensing (DTS).

Keywords: Safe navigation, non-intrusive monitoring of fluid mud, transmission seismic measurements, reflection seismic measurements, yield stress, distributed acoustic sensing (DAS), distributed temperature sensing (DTS)

1. Introduction

Safe navigation through fluid mud is increasingly important because enhancing the navigability with less dredging can help lower transportation costs and benefit biodiversity. The areas with fluid-mud layers need to be routinely surveyed to provide navigation charts used by the vessels. Fluid mud is described as a highly concentrated non-Newtonian suspension of sediment consisting mainly of water, organic matter, silt and clay minerals [1]. Fluid mud is a crucial factor when determining the nautical depth (nautical bottom). It is typically defined by a density value [2]. For example, the Port of Rotterdam uses the density of 1.2 kg/L as a nautical-depth criterium. Other parameters are though also used – for example, the Port of Emden adopts the yield stress of 100 Pa to define the nautical depth [2, 3]. Thus, it is important to have an accurate parameter that description of the fluid mud and could be used in the same way in different ports.

Full-scale and scaled experiments for safe ship navigation in the ports and waterways have been performed already for several decades [3–5]. Traditional ways of characterizing fluid mud involve its sampling, which inevitably disturbs the mud. Other methods, for instance radioactive probes, such as X- and γ-ray tube, can be used to measure the density of the fluid mud, where the density calculation is based on the Lambert–Beer Law [6]. The density profiler based on X-rays – DensX, and the Graviprobe, which measures the cone-penetration resistance and pressures when sinking freely in the water-mud column, can be used to estimate the density and undrained shear strength, respectively [7, 8]. Although these tools can provide a quantitative information about the densities and strength of mud, non-intrusive characterization and monitoring of fluid mud in ports and waterways is preferable. Currently, echo-sounding measurements are used as non-intrusive techniques for assessment of the nautical depth, for which the relationship between the acoustic impedance and densities of the fluid mud are investigated [9]. Multi-beam echo-sounders are deployed to detect fluid-mud layers. Utilization of signals at a higher frequency (200–215 kHz) and at a lower frequency (15–40 kHz) provides an estimate of the approximate thickness of the fluid-mud layer [7]. The higher-frequency measurements are used to map the lutocline, while the lower-frequency measurements provide an estimate of the sea-floor depth. Schrottke and Becker deployed a high-resolution side-scan sonar with a frequency of 330 kHz and a parametric sub-bottom profiler with frequencies of about 100 kHz for detecting the fluid mud with high vertical resolution [10]. The velocimeter, especially the acoustic Doppler velocimeter, was developed on the basis of ultrasonic waves to measure turbulency velocities in the fluid-mud sediments [11]. The mentioned techniques, though, rely on longitudinal (P-) waves, which are related to the bulk properties of the materials.

The propagation velocity and amplitude of transverse (S-) waves strongly depend on the geotechnical properties of the sediment, such as fluid mud [12]. Thus, S-waves could be used to characterize the fluid mud more precisely than when using P-waves and thus bulk properties. However, in seismic exploration in marine environments, the sources and receivers are usually deployed in the water column, more often relatively close to the water surface. Thus, the sources, such as airgun arrays, give rise to P-waves, and the receivers, usually towed by a vessel as streamers, record P-waves as well. This limits the utilization of S-waves because extracting the S-wave information is rather more involving and time-consuming [12]. Still, strong P-to-S-converted waves could be generated at the water bottom, and their utilization for characterization of the fluid mud is possible. A technique that could allow direct extraction of the S-wave velocities is seismic interferometry (SI) for retrieval of non-physical reflections. SI is a method that retrieves new recordings from existing recordings most often by cross-correlation [13–15] of the existing recordings. When the required assumptions for the practical application of SI are not met, non-physical arrivals are also retrieved. Some of the non-physical arrivals arise from internal reflections between layer boundaries [16–18]. SI can thus be applied for targeted retrieval specifically of non-physical (ghost) reflections to estimate the layer-specific velocities for layers in the subsurface [17, 18].

Ultrasonic transmission measurements of marine sediments have been performed, and it was reported that the P-wave attenuation coefficients indicate changes in the sediment composition more distinctly than the velocity of the P-waves [19]. Additionally, relationships between the porosity and P- and S-waves velocities were examined [19]. Leurer [20] carried out pulse-transmission measurements with a center frequency of 50 kHz and reported that in a foraminiferal mud the P-wave velocities range between 1840 m/s and 2462 m/s. Using a center frequency of 100 kHz under different effective pressures, it was also found that the S-wave velocities range between 450 m/s and 975 m/s [20]. Other studies showed that the S-wave velocity in mud samples can be as low as 7 m/s when using signals with a center frequency of 200 Hz [21, 22]. These different values for the P- and S-wave velocities show that it is necessary to perform seismic (ultrasonic) measurements for characterization of the fluid mud for each specific location, i.e., for each port or waterway. This would favor utilization of reflection measurements like in seismic exploration as they can be performed more easily. Additionally, seismic reflection experiments can be conducted with the aid of synthetic seismogram analysis to investigate the shear-wave velocity structure of the shallow-water sediments [23].

Seismic measurements for characterization and monitoring of the subsurface targets are also performed by means of distributed acoustic sensing (DAS) and with distributed temperature sensing (DTS). DAS has already been successfully used in the field of earthquake seismology [24–28], vertical seismic profiling [29, 30], and ambient noise velocity inversions [31]. DTS measurements have been used for monitoring of subsea structures [32] and of carbon capture, utilization and storage [33]. Thus, DAS and DTS could also be very useful in for characterization of fluid mud.

Utilization of DAS and DTS to measure seismic waves in the water and fluid mud offers advantages over the conventional electrical sensors such as electric isolation, immunity to electromagnetic interference, but also that they are non-conductive and non-corrosive, making them well-suited with regard to safety and durability for utilization in liquid-level sensing [34–37]. Such practical advantages are complemented by economical ones. There has been a rapid development in the optical fibers due to their wide usage by the communication industry. This has led to a substantial decrease in price, as well as an increase in performance. For instance, a single-mode optical fiber that used to cost $ 20$/m in 1979 costed just 0.1 $/m in 2008 [38]. Given that the optical fibers are relatively cheap and require little to no maintenance, they could be very useful, from an economical point of view, as receivers for monitoring the nautical depth in ports and waterways. With the experiments we describe below, we investigate the utilization of optical fibers as receivers for fluid-mud level detection and characterization.

In the following, we use laboratory ultrasonic experiments to investigate how P- and S-wave measurements can be used for fluid-mud characterization. We discuss the latest results of seismic (ultrasonic) measurements of P- and S-waves propagation through fluid mud. In Section 2, we first describe the materials, sample preparation, and the rheological experiments for measuring the yield stress. We then introduce the ultrasonic measurements systems we use for transmission and reflection measurements. Subsequently, we describe the DAS and DTS measurement setups.

In Section 3, we present the results from the transmission measurements for monitoring possible changes of the P- and S-wave velocities when the ultrasonic signals propagate through fluid mud at different stages of consolidation. We link the observed transmission velocity changes to the measured yield stress during the same consolidation stages of the fluid mud. Further, we describe results from the reflection setup for estimating the layer-specific P- and S-wave velocities of the fluid mud. Finally, we validate the utilization of DAS and DTS as seismic and temperature receivers in laboratory experiments for detecting the fluid-mud/water interface.

In Section 4, we discuss the accuracy of our results and their applicability to other ports, while in Section 5 we draw conclusions.

2. Characterization and monitoring of fluid mud in a laboratory

We develop laboratory ultrasonic measurement systems for transmission and reflection seismic measurements for characterization and monitoring of fluid mud while it is consolidating. The transmission seismic-measurements systems are designed for direct, fast, point-to-point measurements in the fluid mud using ultra-sonic transducers or DAS as receivers. The reflection seismic-measurements system uses ultrasonic transducers to record waves that have reflected or refracted at different layer boundaries including the bottom of the water layer and the bottom of the fluid-mud layer. The reflection measurements can be used to record common-source gathers, which can subsequently be utilized to characterize veloc-ity changes in the fluid mud during the consolidation using seismic-exploration techniques. We also describe the laboratory setup for rheological measurements of the fluid mud and the setup for DTS measurements.

2.1 Fluid-mud sample preparation and handling

For the transmission and reflection measurements, we use fluid-mud samples extracted from the Calandkanaal (Port of Rotterdam) at the location indicated in **Figure 1a**. Before conducting the measurements, we stir a sample using a mechan-ical mixer in order to obtain a homogeneous volume of fluid mud with a uniform density. The density of the homogenized sample is 1197 kg/m3. After the homoge-nization, the fluid-mud sample appears like a mud slurry (**Figure 1b**). The samples are consecutively left to consolidate through a self-weight process. We perform ultrasonic measurements while the fluid mud is consolidating. Synchronously with the ultrasonic measurements, we also perform rheological measurements to inves-tigate the yield stress. We investigate the fluidic yield stress using a recently

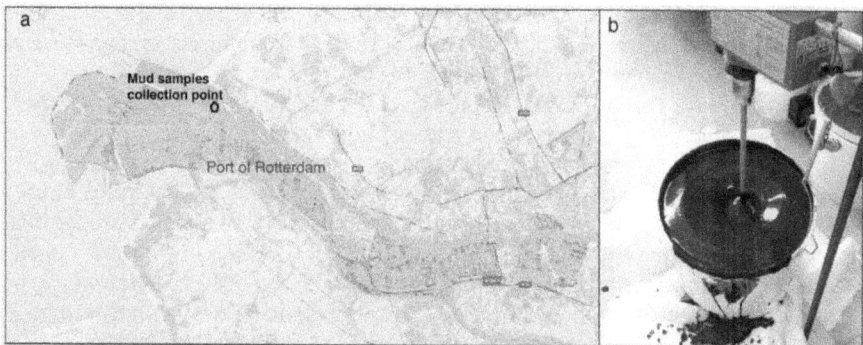

Figure 1.
(a) Map of the port of Rotterdam illustrating the location of the site from where the fluid-mud samples had been collected (source: Google maps). (b) The process of homogenizing fluid mud with a mechanical mixer.

developed protocol for the fluid mud [39, 40]. We use a HAAKE MARS I rheometer (Thermo Scientific) with two measuring geometries (Couette and vane) and apply stress ramp-up tests to measure the yield stress. The stress ramp-up tests are performed using a stress increase from 0 to 500 Pa at a rate of 1 Pa/s, until the shear rate reaches 300 s^{-1}, under a stress-control mode.

2.2 Transmission seismic measurements with transducer receivers

The transmission seismic laboratory setup is equipped with two pairs of piezo-electric ultrasonic transducers (**Figure 2b** and **c**). Each pair consists of a source and receiver transducer, with one of the pairs using P-wave transducers and the other pair – S-wave transducers. The direct transmission measurement represents a point-to-point measurement with both transducer pairs placed along the horizontal direction. Because of this source-receiver geometry, the estimated velocities of the P- and S- waves correspond to transmissions along horizontal layers inside the fluid mud, if such layers are developed.

As shown in **Figure 2a**, the laboratory setup includes a fluid-mud tank, a signal-control part, and the two pairs of ultrasound transducers. The signal-control part in turn consists of a source-control part and a receiver-control part. In the source-control part, a function generator produces a desired signal, which signal is subsequently passed to a power amplifier to be finally passed to the source transducer, which sends it through the fluid mud. The fluid-mud tank is a plastic box that has

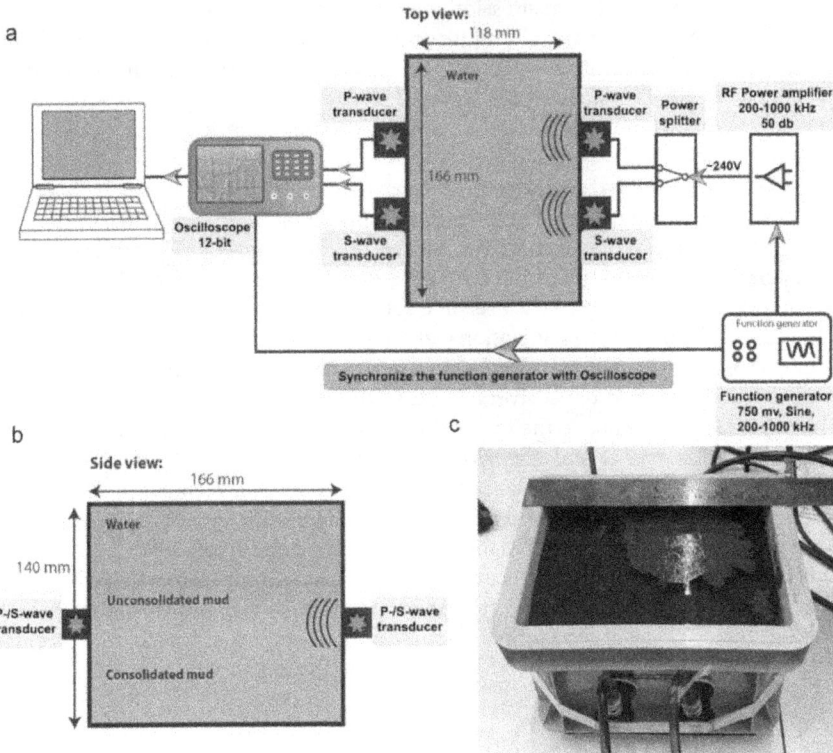

Figure 2.
(a) Sketch of the transmission seismic laboratory setup with the fluid-mud box viewed from above and showing the horizontal arrangement of the two transducer pairs. (b) Side view of the fluid-mud box showing the vertical alignment of the ultrasonic transducers. (c) Photo of the fluid-mud box showing also the two source transducers.

opening for the installation of the transducer end-caps. The receiver-control part of the setup consists of the receiver transducers, attached to the fluid-mud tank using end-caps, an oscilloscope for digitalization and displaying, and a computer, connected to the oscilloscope, to record the sensed signals. The generated source signal is also visualized on the oscilloscope for quality control.

For the transmission measurements, we use as a source signal a gated sine-wave pulse with a center frequency of 1 MHz. A measurement is performed using a pulse-time delay. To increase the signal-to-noise ratio, especially needed for the S-wave velocity estimations, a measurement at each stage of consolidation consists of 1024 repeated recordings summed together to obtain a final transmission recording. This is done for both the P- and S-wave pair.

For each stage of the measurements, the first step in estimating the propagation velocities is to pick the first arrivals of the P- and S-waves. The second step is to calculate the P- and S-wave velocities by dividing the travel distance of waves, which is the distance from the source to the receiver transducer within each pair (equal for both pairs), by the travel times estimated from the picked first arrivals.

2.3 Reflection seismic measurements with transducer receivers

Similar to the transmission seismic laboratory setup, the reflection system consists of a signal-control part, a fluid-mud tank, and ultrasound transducers, but further to that also includes a transducer-placement part (**Figure 3**). While the signal-control part is the same as for the transmission measurements (**Figure 3b**), the fluid-mud tank is different and only one pair of ultrasonic transducers is used (**Figure 3c**). The transducer-placement part allows changing the positions of the transducers by moving them along horizontal and vertical bars (**Figure 3a** and **c**). This facilitates recording of reflections at multiple horizontal positions to obtain reflection common-source gathers, if desired with sources and receivers at different depths.

In the measurements we perform, the transducers are placed a certain distance above the top of the fluid-mud layer to better mimic a geometry of a marine seismic-exploration survey. While placing sources and receivers during a field measurement campaign directly at the top of the fluid mud would allow direct recording of S-waves, a recording geometry with seismic sources and receivers towed at a certain height above the bed in the navigational channel is more practical – the surface of the sediments is seldomly flat, and hard object protruding from the sediments could damage the sources and/or receivers. On the other hand, towing the sources and receivers at a distance above the top of the fluid-mud layer inevitably brings uncertainty in the estimated seismic velocities caused by the salinity and temperature of the water. It is possible to monitor the changes in the salinity and temperature at specific locations, but the uncertainty still remains when using such point measurements for larger-area surveys due to the dynamics of the marine environments.

In order to eliminate these uncertainties, we apply SI for retrieval of ghost reflections from inside the fluid-mud layer and eliminate the travel-paths of the waves in the water layer. For pressure measurements in water, like in our laboratory setup, a general representation of SI by cross-correlation is [41].

$$p(R2, R1, t) + p(R2, R1, -t) \propto \sum_{S=S1}^{SN} p(R2, S, t) \otimes p(R1, S, t), \qquad (1)$$

where $p(R2, R1, t)$ is the retrieved pressure response at receiver at $R2$ from a virtual source at the position of a receiver at $R1$, $p(R2, S, t)$ is the pressure response measured at $R2$ from a source at S, with $S1 \ldots SN$ sources distributed evenly over

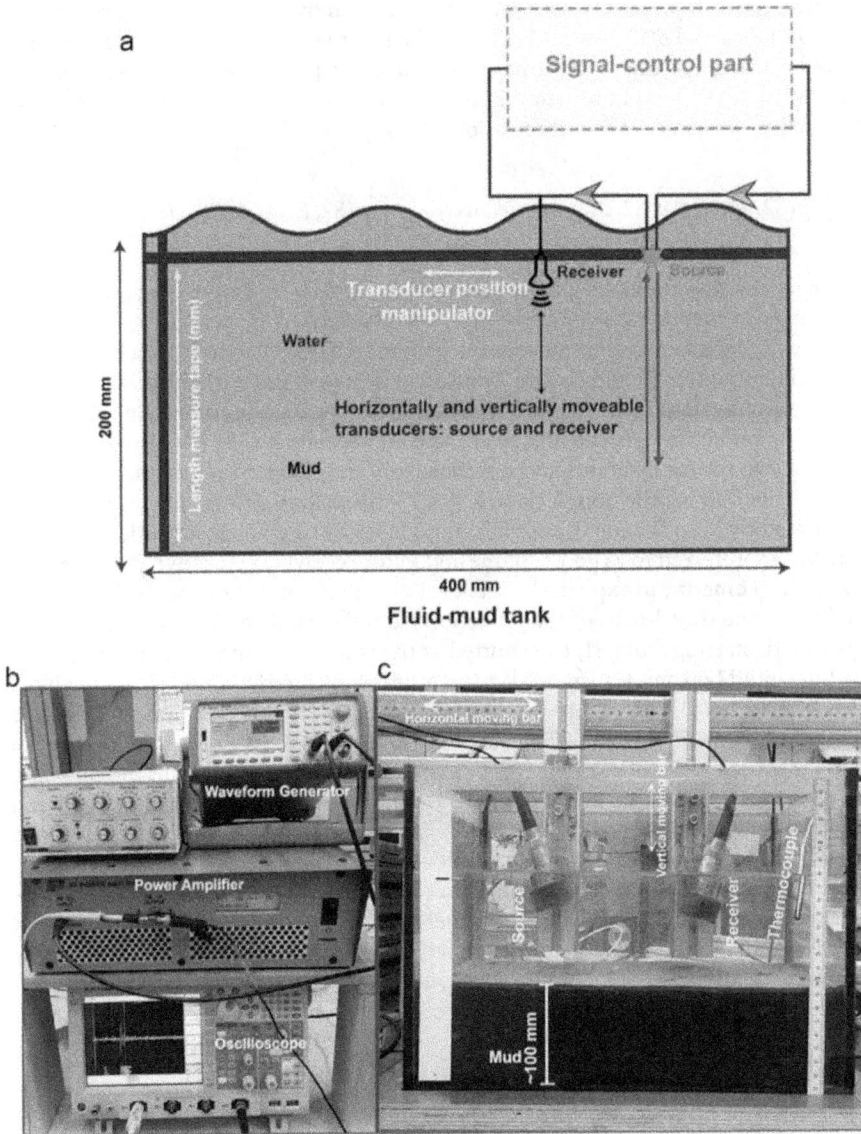

Figure 3.
Reflection seismic-measurements system. (a) Cartoon of the fluid-mud tank with the transducer-placement part and the signal-control part (identical to the one in the transmission measurements). Red star indicates the source and black probe indicates the receiver. The transducer-placement part allows vertical (blue arrows) and horizontal (white arrows) displacement of the source and receiver. (b) Photo of the signal-control part. (c) Photo of the fluid-mud tank and the transducer-placement part with a source and receiver ultrasonic transducers.

surface that effectively surrounds the two receivers, $-t$ indicates time reversal (acausal time), and $\otimes$ indicates correlation. As mentioned above, when the assumptions for this simplified representation are not met [14], e.g. as in a seismic reflection survey when the sources are only at the surface and thus do not surround the receivers, ghost reflections are retrieved [17, 18], and we can write

$$p(R2, R1, t) + p(R2, R1, -t) + ghosts \propto \sum_{S=SK1}^{SK2} p(R2, S, t) \otimes p(R1, S, t), \qquad (2)$$

where *ghosts* represents retrieved non-physical arrivals, including ghost reflections, and $SK1$ and $SK2$ now indicate that the summation is only over sources on a limited surface. For practical purposes, in our laboratory setup we choose to have only two source positions and multiple receiver positions. Using source-receiver reciprocity, we can thus rewrite relation (2) as

$$p(S2, S1, t) + p(S2, S1, -t) + ghosts \propto \sum_{R=RK1}^{RK2} p(S2, R, t) \otimes p(S1, R, t), \qquad (3)$$

where the summation is now over receiver positions and we retrieve a pressure recording at a virtual receiver at the position of source $S2$ from a sourse at $S1$. Thus, to apply SI, we use two common-source gathers (CSGs). The two source positions (labeled Source 1 or S1 and Source 2 or S2 in **Figures 4** and **5,** respectively) at the same height and distanced in the horizontal direction 50 mm from each other. We record the reflected signal at a receiver, labeled Receiver 1 (R1) in **Figure 4,** aligned with the two source positions and distanced 100 mm from S1 (and thus 50 mm from S2). Following the nomenclature in [17], the source and virtual receiver redatumed by SI to the top of the fluid-mud layer during the ghost-reflection retrieval are referred to as ghost source and ghost receiver, respectively. Assuming a favorable geometry, to explain the retrieval of a ghost reflection inside the fluid-mud layer, the travel-path of the reflection from the fluid-mud bottom, i.e., the travel-path starting from S1, transmitted at the water/mud interface, reflected by the fluid-mud bottom, transmitted at the mud/water interface, and then arriving at R1 is labeled 1–2–3-4 in **Figure 4.** The travel-path of the reflection from the water/mud interface, starting from S2 and arriving at R1, is labeled 1′-4′. Cross-correlation of the recorded reflections at R1 from S1 and S2 will effectively result in removal of the common travel-paths in 1–2–3-4 and 1′-4′. Thus, the parallel travel-paths 1 and 1′ and the coinciding travel-paths 4 and 4′ are eliminated, and only the travel-path 2–3 is left over representing a ghost reflection only inside the fluid-mud layer from a ghost source and ghost receiver placed directly at its top (**Figure 4**). In reality, the exact receiver position ensuring that travel-paths 1 = 1′ and 4 = 4′ is unknown. Because of that, recordings at multiple receiver positions from both sources are required, i.e., two CSGs. To obtain such gathers, we displace the receiver from position R1 to the right along the horizontal bar by 5 mm multiple times and record

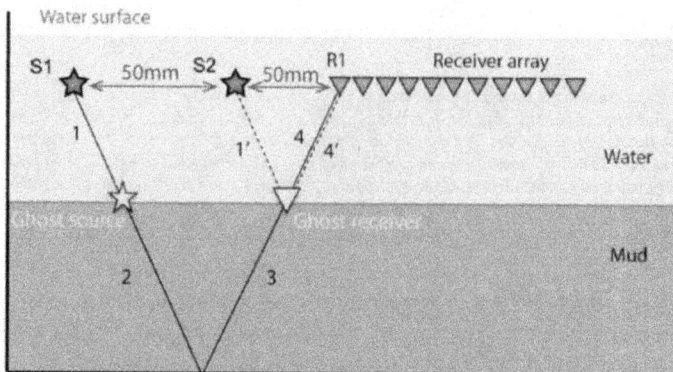

Figure 4.
Illustration of the geometry needed for retrieval of ghost reflections from inside the fluid-mud layer. See text for explanation of the symbols.

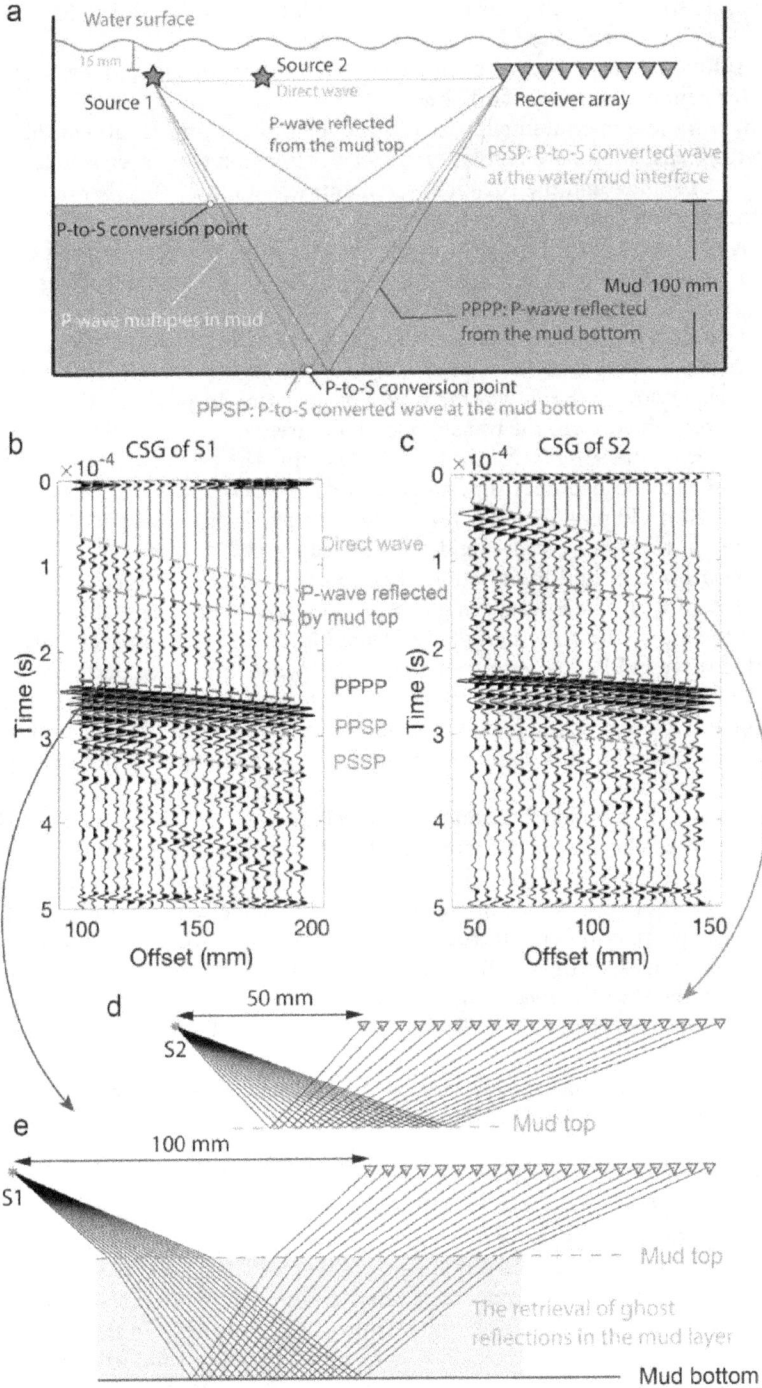

Figure 5.
(a) Illustration of the travel-paths of the expected arrivals from S1 to a receiver in the reflection measurements. (b) Wiggle plot of the recorded CSG from S1. (c) Wiggle plot of the recorded CSG from S2. (d) Sketch of the travel-paths of the primary reflections of the mud top in the CSG from S2. (e) Sketch of the travel-paths of the primary reflection of the mud bottom (PPPP) in the CSG from S1. The ghost reflection is retrieved by summing the individual arrivals highlighted in green in (e) obtained from cross-correlating the primary reflection from the fluid-mud top in the CSG from S2 with the primary reflection PPPP in the CSG from S1.

for the same source at each receiver position. In this case, we record at 20 positions. That is, the CSGs for S1 and S2 consist of 20 traces each.

The source signal we use is similar to the one for the transmission measurements but with a center frequency of 100 kHz.

Also with these measurements, to increase the signal-to-noise ratio of the recorded signals, a measurement at each receiver position from each source is repeated 1024 times and the 1024 measurements are summed together to obtain a final trace for that source and receiver positions.

Using the travel-path sketches in **Figure 5a**, we explain several arrivals of interest in the CSGs. **Figure 5b** and **c** present wiggle plots of the recorded CSGs from S1 and S2, respectively. We calculate expected arrival times based on the source/receiver offsets and the thicknesses of the water and fluid-mud layers, each of which we can directly measure. For propagation through the water layer, we use P-wave velocity of 1500 m/s. For the waves propagating through the fluid mud, we use values estimated from the transmission measurements – 1570 m/s for the P-wave velocity and 958 m/s for the S-wave velocity. The calculated reference times are illustrated by dashed lines superimposed on the CSGs to assist in interpretation of the arrivals. In **Figure 5b** and **c**, the reflection arrivals of interest in this study are the primary reflection from the fluid-mud top (magenta) and the three primary reflections from the fluid-mud bottom that are labeled as PPPP (blue), PPSP (red), and PSSP (orange). The S-waves in the experiment appear as waves converted from P to S at the top or the bottom of the fluid-mud layer. For example, the P-to-S converted wave in PPSP is generated when the P-wave impinging on the fluid-mud bottom is reflected as an S-wave; the P-to-S converted wave in PSSP is generated when the P-wave impinging on the fluid-mud top in transmitted to the fluid mud as an S-wave (**Figure 5a**) and continues to propagate as an S-wave until reaching the fluid-mud top again.

To retrieve ghost reflections, one can use relation (3) and correlate the CSGs. Such an approach could result in other retrieved arrivals interfering with the desired ghost reflections. To avoid that, we follow [17] and correlate only specific arrivals. To retrieve a P-wave ghost reflection from inside the fluid mud, we cross-correlate the primary reflection from the fluid-mud top in the CSG from S2 (**Figure 5d**) with the primary reflection PPPP in the CSG from S1 (**Figure 5e**). In a similar way, the P-to-S converted ghost reflection and S-wave ghost reflection are retrieved using the reflections PPSP and PSSP in the CSG from S1, respectively.

2.4 Transmission seismic measurements using DAS and measurements with DTS

We use a standard single-mode communication fiber for both the DAS and DTS measurements. This means that we can combine the two methods and compare the difference in their performance With DAS, such fibers can act as seismic receivers that measure the dynamics of a strain field acting on a fiber [42]. With DTS, such fibers can act as strain and temperature sensors (and thus also labeled DT(S)S), which measure the static strain and temperature along the fiber [43].

To verify that these fibers can serve as receivers for fluid-mud level detection and characterization, we conduct seismic and temperature laboratory experiments using commercially available interrogators. These interrogators are the iDAS from Silixa and DITEST STA-R from Omnisens for measuring the acoustic impedance and temperature, respectively. For a more detailed explanation of the iDAS system, the reader is referred to [42].

Our fiber is coiled around a PVC pipe with a diameter of 0.125 m, which allows us to use more fiber and, hence, have more measuring points than when using a straight fiber. In addition, the coining increases the vertical resolution by compressing the gauge length of 10 m of the cable (the length over which the back-scattered signal is averaged to increase the signal-to-noise ratio of the detected dynamic deformation) only over a few vertical centimeters. Due to the coiling, we also change the directional sensitivity [44], making the cable more sensitive to horizontal waves, with respect to the column. The PVC pipe with the fiber coiled on it is placed inside a transparent column. We first perform experiments with two types of synthetic clay, namely kaolinite and bentonite, and subsequently with two types of fluid mud – one from the Port of Rotterdam, which is the same sample mud as described above, and the other from the Port of Hamburg. For the experiments with the synthetic clays, we fill the lowest part of the column, without coiled optical fiber, with sand. Above the sand, we put one of the clays, and then we fill the remainder with water. For the fluid-mud experiments, we instrument also the lowest part of the column with fiber and start filling the column with one of the fluid muds starting already at the bottom, while we again fill the remainder of the column with water. A schematic overview and pictures of the setup are shown in **Figure 6**. Note that for the measurements with kaolinite and bentonite, we have 0.5 m in depth, which is 123 m in fiber length, acting as sensors. For the measurements in the muds, we added 0.2 m in depth, giving us a total of 171 m of fiber length, acting as sensors. For both setups, we have 10 m of fiber outside of our column to use as a reference.

With DAS, we try to capture the water/mud interface and measure the shear strength build-up. We test various sources for these purposes. Our sources include a small transducer with a center frequency of 500 kHz, a larger transducer with a center frequency of 200 kHz (**Figure 6b** and **c**) and a common duo echo-sounder with a center frequency of 38 kHz and 200 kHz, which is also used by marine vessels to measure depth. We connect these sources to the same source-side signal-control part as described above. We use a frequency range from 25 kHz to 45 kHz, since preliminary results indicated that this range should give the best results. The sampling frequency of the DAS system is set at the maximum of the system, which is 100 kHz.

For the DTS measurements, we use two standard heating rods, which we place 5 cm away from the fiber, to heat the column and measure the difference with respect to time along the column. This we only do for the kaolinite sample, since a very similar result is expected for the other clay and two mud samples.

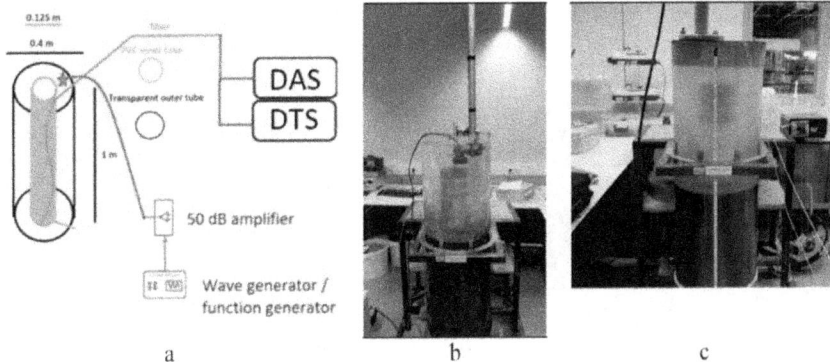

a   b   c

Figure 6.
(a) Schematic overview of the setup for DAS and DTS measurements. A photo of the column with the optical fiber wound around the PVC pipe when using mud from the (b) port of Rotterdam and (c) port of Hamburg.

3. Results

We describe the results of the ultrasonic transmission measurements with ultrasonic transducers and correlate them to the results from the rheological measurements. We further report the results from the reflection measurements and how they were used to retrieve ghost reflections. We then show the results from the DAS and DTS measurements.

3.1 P- and S-waves velocities in the fluid mud from transmission measurements with ultrasonic transducers

We examine the first arrivals of transmitted P- and S-waves and estimate their velocity variations during the consolidation of the fluid mud. We do not observe a detectable change in the P-wave velocity – the P-wave first arrivals appear to be constant throughout the consolidation process (**Figure 7a**). This finding agrees with previous results reporting that the S-wave velocity is more sensitive to changes in lithology and mechanical properties than the P-wave velocity [45]. The traveltime of the direct arrivals of the P-wave is 0.074 ms (**Figure 7a**), and thus the corresponding velocity is 1570 m/s. By examining the change in arrival time of the

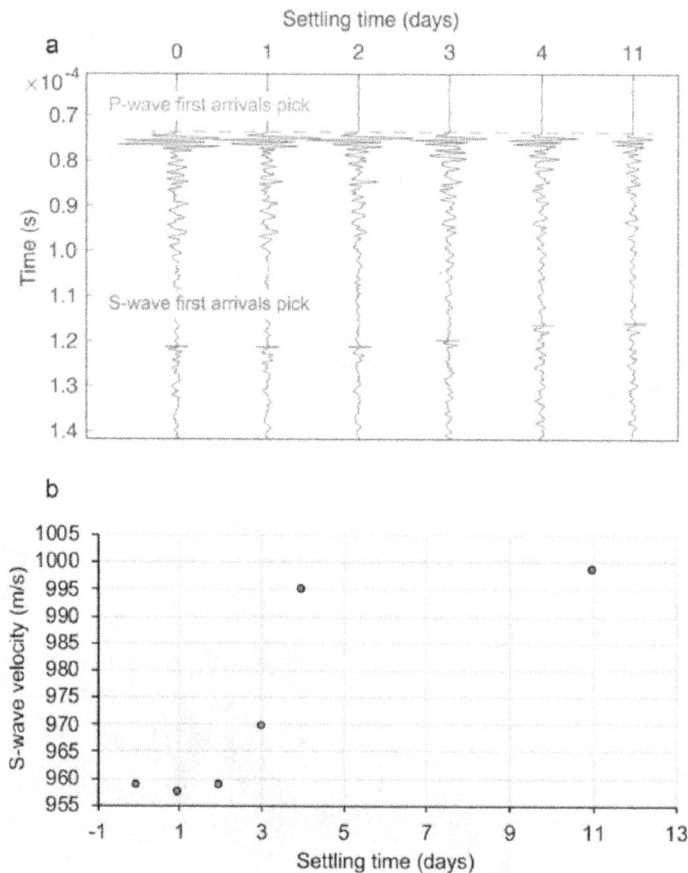

Figure 7.
(a) Transmission recordings of the direct P- and S-wave arrivals as a function of consolidation time. (b) Estimated S-wave velocity as a function of the consolidation time.

first S-wave arrival (**Figure 7a**), we find that the S-wave traveltime decreases with consolidation time, indicating that the S-wave velocity increases with the consolidation progress (**Figure 7b**).

We can see from **Figure 7**, that during the first three days the S-wave velocity is nearly stable exhibiting very little fluctuations. Starting from Day 3, the S-wave velocity shows a strong increase from 959 to 995 m/s during the next two days. In the second week, the S-wave velocity only experiences a small increase and eventually reaches 998 m/s. By comparing the velocity variations of the P-wave and S-waves, we can summarize that the relative increase in the S-wave velocity is much stronger than in P-wave velocities, validating the statement that the S-waves are much more sensitive to the consolidation of the fluid mud than the P-waves. This finding agrees with a previous in-situ seismic exploration results using pulse-transmission techniques [45].

By drawing the estimated S-wave velocities from **Figure 7b** as a function of the concurrently estimated fluidic yield stresses (**Figure 8**), we see a positive correlation during the consolidation of the fluid mud. The correlation appears to indicate that the S-wave velocity starts increasing after the fluidic yield stress exceeds some critical value (for each of the Couette and vane geometry). Once the critical value is surpassed, the S-wave velocity increases with the increasing fluidic yield stress caused by the consolidation.

3.2 P- and S-wave velocities inside the fluid-mud layer from ghost reflections

The recorded primary reflections from the fluid-mud top and bottom are identified and shown in **Figure 9**. We apply SI using the reflection from the mud top in the CSG from S2 and the primary reflections PPPP, PPSP, and PSSP from the mud bottom in the CSG from S1 (**Figure 9**). As explained in Section 2.2, the ghost reflections are retrieved by eliminating the P-wave travel-paths inside the water. The ghost reflections in **Figure 10**, retrieved using the primary reflections PPPP, PPSP, and PSSP, are labeled PP, PS, and SS, respectively. In **Figure 10**, we also show the length of each of the legs of the reflection travel-paths of the retrieved ghost reflections. We use these lengths to estimate the wave velocities using the arrival times of the retrieved ghost reflections.

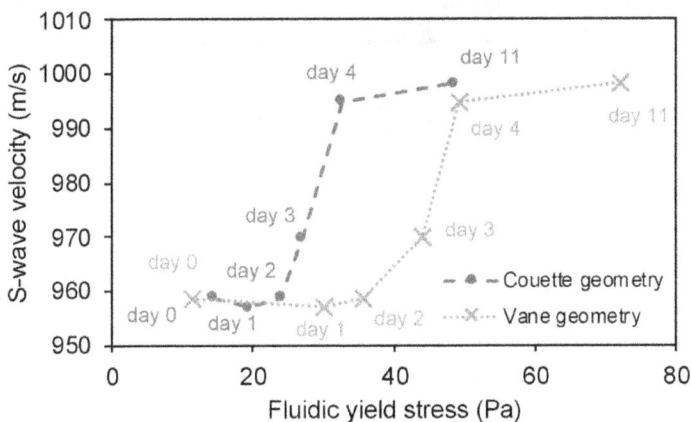

Figure 8.
*Relationship between the estimated S-wave velocities (**Figure 7b**) and the concurrently estimated fluidic yield stress, using Couette and vane geometry, as a function of the consolidation time.*

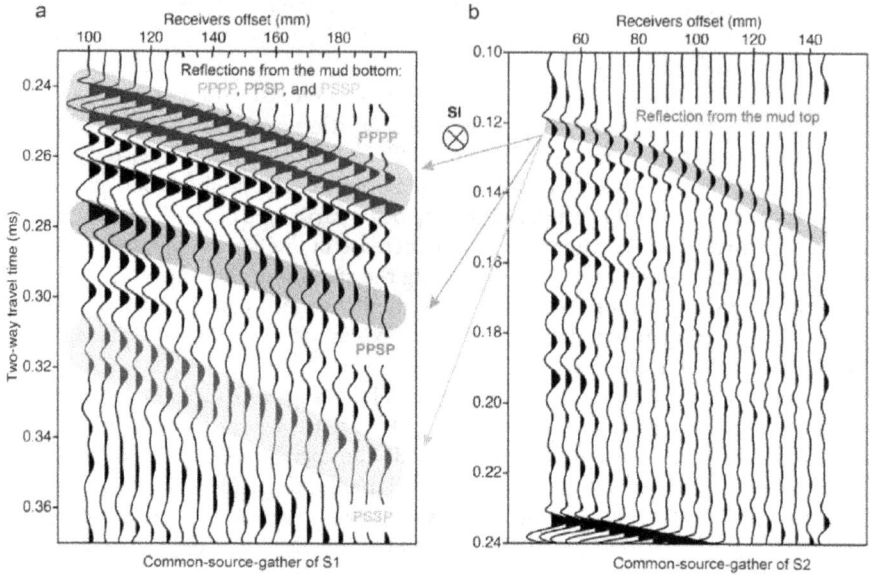

Figure 9.
Identified primary reflections in the common-source gather from (a) source 1 and (b) source 2. We apply seismic interferometry (SI) by correlating (the ⊗ symbol) the reflection from the mud top with each of the three identified reflections from the mud bottom followed by summation over the receivers (Eq. 3).

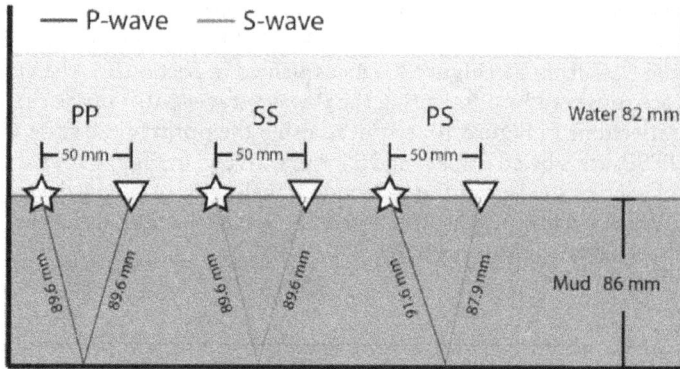

Figure 10.
The travel distances of the travel-paths of the ghost reflections PP, SS, and PS when the fluid-mud thickness is 86 mm, which is the thickness on day 11 of the consolidation.

As explained, the retrieved result is obtained by stacking the correlated traces. When the receiver array is sufficiently long, the stacking would have resulted in the retrieved ghost reflections only, with the contribution to the retrieved signal coming from summation inside the so-called stationary-phase region [46], i.e., the region where a curve appears nearly horizontal. In **Figures 11a–13a**, we indicate the stationary-phase regions with green dashed rectangles. Because our receiver array is of a limited length and is further only on one side of the sources, summation of all traces produces more or less erroneous results (**Figures 11b–13b**). Because of this, to retrieve the ghost reflections we use for the summation only traces in the stationary-phase region (**Figures 11c–13c**). We then pick from those results the two-way traveltimes to estimate the velocities inside the fluid-mud layer.

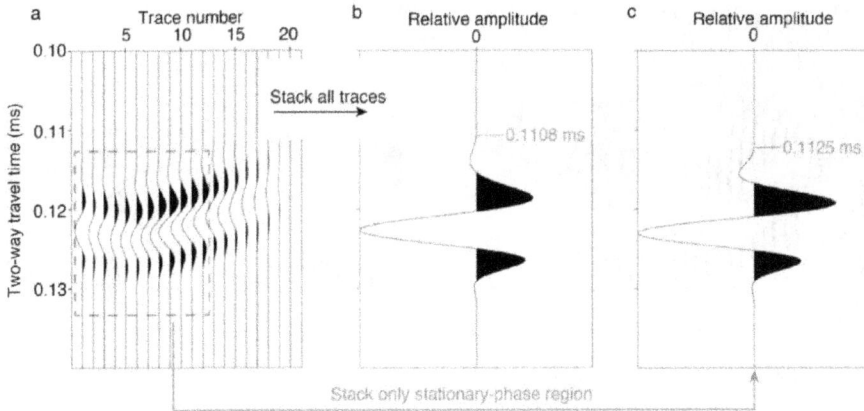

Figure 11.
*Two-way traveltime pick of the ghost reflection PP. (a) Correlation result of the reflection from the fluid-mud top from **Figure 9b** with the PPPP reflection from **Figure 9a**. The stationary-phase region is indicated by the dashed green rectangle. The retrieved ghost reflection PP when summing over (b) all traces in a and (c) the traces inside the stationary-phase region in a.*

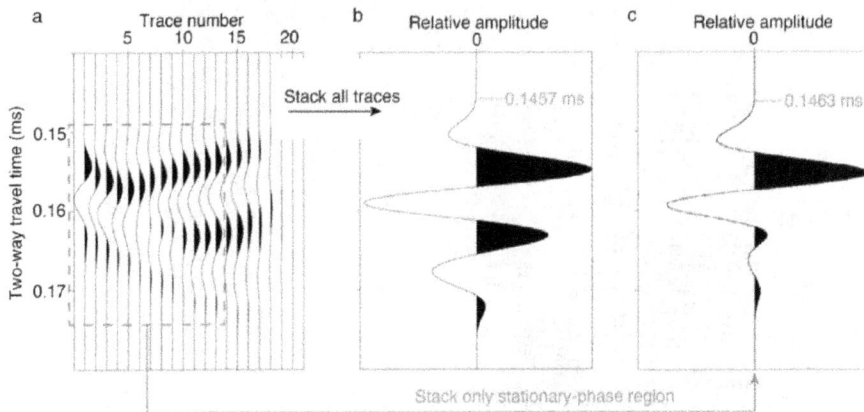

Figure 12.
*As in **Figure 11** but for ghost reflection PS. The correlation in (a) is with the PPSP reflection.*

Dividing the travel distance of 179.2 mm, which ghost reflection PP has traversed inside the fluid-mud layer (**Figure 10**) by the picked two-way traveltime from **Figure 11c**, we estimate the P-wave velocity to be 1592 m/s. To estimate directly the S-wave velocity inside the fluid-mud layer, we divide the travel distance the ghost reflection SS has traversed inside the fluid-mud layer, again 179.2 mm (**Figure 10**), by the picked two-way traveltime from **Figure 13c**, and obtain 995 m/s. Comparing this value with the estimated value from the transmission measurements on day 11 of 998 m/s (**Figure 7b**), we see that the difference is only 0.3%, which is negligible. Comparison of the estimated P-wave velocity to the value from the transmission measurements of 1570 m/s, we see that the difference is 1.4%, which is a bit higher but still acceptable.

3.3 Detection of the water/mud interface using DAS and DTS

Figure 14 shows DAS measurements of the arrivals recorded along the fiber as a function of arrival time when using the fluid mud from the Port of Hamburg and

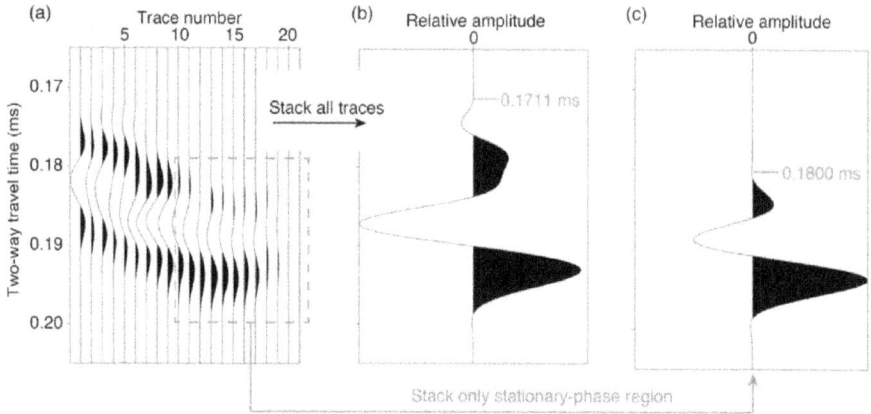

Figure 13.
*As in **Figure 11** but for ghost reflection SS. The correlation in (a) is with the PSSP reflection.*

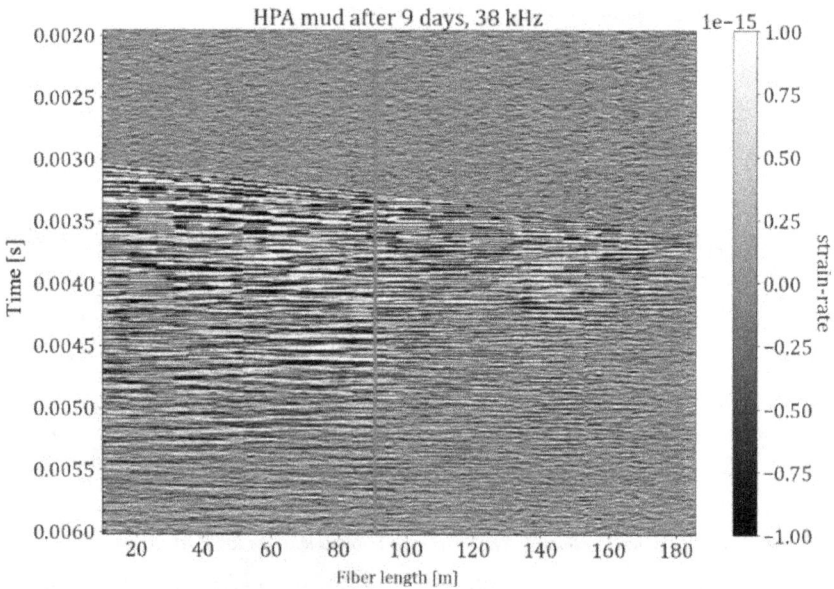

Figure 14.
*DAS recordings using the setup from **Figure 6c** showing direct arrivals and multiple reflections. The blue line indicates the water/mud interface, which is at 90.7 m along the fiber.*

the large transducer as a source (**Figure 6c**). We perform the measurements after the mud has consolidated for 9 days. To improve the signal-to-noise ratio, we repeat the recordings 10 times and then stack them. Using the first arrivals, i.e., the direct P-wave, we estimate the P-wave velocity in water to be around 1450–1500 m/s, while in the fluid mud to be 1490–1570 m/s. The reason for the uncertainty is likely related to the relatively low rate of time sampling of 100 kHz for the source frequency we use of 25–45 kHz. For this sampling rate, the Nyquist frequency is 50 kHz, which is very close to the source frequencies and, thus, makes the velocity analysis more ambiguous. The small difference in the P-wave velocity of the water and the fluid mud combined with the uncertainties make the detection of the water/ mud interface rather challenging if the first arrival as used.

The recordings in **Figure 14** show that a more accurate and robust criterion to detect the water/mud interface is to look at the multiple reflections and their amplitude attenuation. Looking at the figure, we can see that later arrivals appear to faint, i.e., are more attenuated after the water/mud interface, with the latter indicated by the blue line. Taking a closer look at the multiple reflections, we see that these later arrivals have completely fainted after 93.7 m fiber length, with the water/mud interface at 90.7 m fiber length. This difference of 3 m of fiber might be related to the gauge length of the fiber, i.e., the length over which the DAS system averages the observations, which in our case is 10 m. Another reason could be the uncertainty in the exact position of the fiber.

The measurements with the fluid mud from the Port of Rotterdam and the two clays show similar results.

We also look at the signal attenuation due to the consolidation of the mud, and thus the increase of its shear strength. **Figure 15a** and **b** show the DAS recordings in bentonite clay performed on the first and second day of the consolidation, respectively. We see a clear difference in signal penetration through the bentonite clay – on the first day, there is little to no signal penetration, opposed to the second day, when the waves propagate all the way through the column. This difference is purely related to the buildup of shear strength in the bentonite, since bentonite does not settle but builds up shear strength with time.

From the tests we perform with different types of sources (small and large transducer and duo echo sounder) we observe that the small transducer with resonant frequency of 500 kHz does not generate enough energy when we use it for emitting a P-wave at 25 kHz – 45 kHz. For that reason, it is outperformed by the big transducer whose resonant frequency of 200 kHz is closer to our target source-signal frequency of 25 kHz – 45 kHz. The duo echo sounder generated by far the strongest signal; however, because it was mounted on the transparent outer column and was situated right above our PVC pipe, a lot of tube waves and refracted waves are generated, which are undesired in our tests. These strong interfering events could potentially be suppressed applying further signal processing, as we suggest above – for example using a frequency-wavenumber filter.

Besides using the optical fiber as a receiver for seismic waves, we also use it as DTS recorder to measure temperature. Due to the difference in the heat capacity and heat conductivity between water and mud, a difference in heating occurs when we start heating up the column using heating rods in the water and kaolinite. This difference can be observed in **Figure 16**, where we show the measured Brillouin

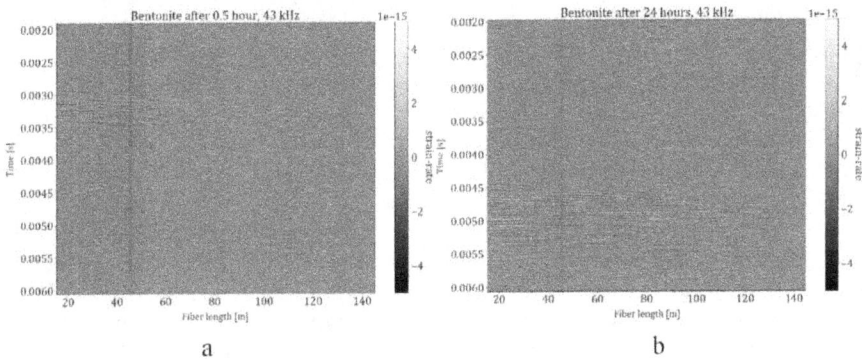

Figure 15.
DAS recordings when synthetic clay (bentonite) is used as fluid mud. The recordings were done after (a) half an hour and (b) 24 hours of consolidation of the bentonite.

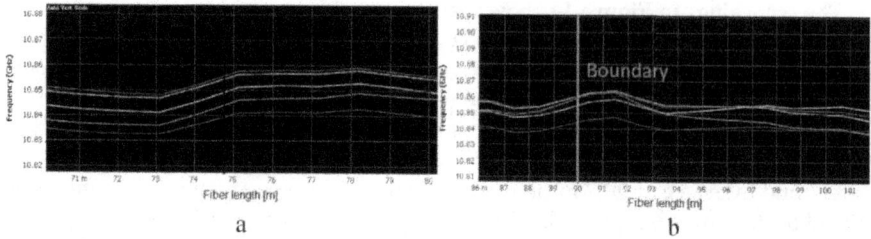

a   b

Figure 16.
Brillouin frequency changes in (a) water and (b) mud after increasing the temperature of the water in the column by 1°C. the red line indicates the water/mud interface. The brown curve represents a reference measurement without heating. The curves with colors other than brown represent measurements after heating up the water several times by 1°C.

frequency when we heat up the water and kaolinite. The brown curves show a reference measurement before the heating, while the other colored curves show the measurements after increasing the temperature of the water each time by 1°C. Inside the water layer (**Figure 16a**), we observe a linear increase in the Brillouin frequency per °C. Inside the mud layer, however, we see a non-linear trend due to the lower heat capacity and lower heat conductivity. This is especially visible along the red curve, which characterizes the first measurement after we start heating up the column: we see that in the lower part, starting at 99 m, the red curve overlaps the brown curve meaning that the heat from the heating rods has not yet reached the fiber at that level and deeper.

The DTS measurements show that interpretation of the water/mud interface can be achieved with a likely accuracy of around 4 cm.

4. Discussion

The direct transmission measurements of the P- and S-wave velocities inside the fluid-mud layer showed that the P-wave velocity is nearly independent of the consolidation process while the S-wave velocity significantly increases during the consolidation. This can be attributed to the property changes of the fluid mud due to the compaction effect of the consolidation and potentially the production of gas in the mud. The S-wave velocity is principally determined by the grain structure and shear modulus of the frame of the solid phase (minerals). The P-wave velocities on the other hand depend on the elastic moduli of the grains, sediment frame, and bulk modulus of the fluid. Thus, for marine sediments with high porosity, such as the fluid mud, the S-wave rather than the P-wave is strongly affected by the consolidation, and, thus, can be potentially used to characterize the consolidation process.

Using SI for retrieval of ghost reflections inside the fluid-mud layer, we removed the kinematic influence of the water layer above the mud. The estimated velocities of the P- and S-waves using the ghost reflections PP and SS, respectively, were very close to the ones estimated from the direct transmission measurements inside the fluid-mud layer. Because we also had the ghost reflection PS (**Figure 12**), we could estimate the S-wave velocity inside the fluid mud also from this arrival. We did this making use of the already estimated P-wave velocity for the propagation along the P-wave path of 91.6 mm in **Figure 10**. The value we then obtained was 991 m/s, which is quite close to the value of the S-wave velocity obtained from the ghost reflection SS, but is of course inheriting errors from the estimation of the P-wave velocity. Nevertheless, all three values can be used as quality control of each other or as substitutes when one of the three ghost reflections cannot be reliably retrieved due to, for example, interference from other arrivals.

Observing the multiple reflections in the DAS recordings, we estimated an error of 3 m along the coiled fiber in detecting the depth of the water/mud interface. Since we coiled the fiber around a PVC pipe with a diameter of 0.125 m and because the fiber's thickness is 1.6 mm, the 3-meter error of fiber length translates to 1.2 cm of vertical error in the depth of the water/mud interface. With such an error, to the best of our knowledge, our approach is the most accurate non-intrusive method for determining the depth of the water/mud interface. Note that to achieve this accurate result, the only processing we applied was to increase the signal-to-noise ratio by the summation of the 10 separate recordings. More signal processing could further improve the determination of the water/mud interface. We expect that a similar high accuracy is achievable in the field as well since the upper end of the optical fiber is placed at the very bottom of the water layer, which limits errors caused by differences in, for instance, the water temperature.

The direct transmission measurements with DAS, on the other hand, allowed estimation of the P-wave velocity in the fluid mud in the range 1490–1570 m/s. Comparing these values to the value of 1570 m/s from the direct transmission measurements horizontally inside the fluid mud means an uncertainty of about 5.1%, which is not negligible. This confirms the difficulty when using a source in the water and receivers in the fluid mud, and clearly underlines the advantage of using SI with ghost reflections from reflection measurements. Thus, we argue that another very useful application of DAS could be with direct transmission measurements inside the fluid-mud layer, and thus also for transmission tomography between a vertical array of sources inside the mud and a vertical DAS pole with coiled fiber.

For our laboratory measurements, we used fluid-mud samples from the Port of Rotterdam and the Port of Hamburg. Nevertheless, our results and conclusions can be generalized to fluid-mud samples from other ports. Because the estimated P- and S-wave velocities using the ghost reflections do not depend kinematically on the water layer, this technique could easily be applied to any port or waterway. Of course, the P- and S-wave velocities of the fluid mud will differ from place to place, so those will need to be estimated for each place, for example for correlation with the yield stress. The DAS and DTS techniques for estimating the water/mud boundary can likewise be used at any other port or waterway, as they depend only on the strong contrast in the observed parameters between the layer and fluid-mud layer.

5. Conclusions

We presented recent results for non-intrusive characterization and monitoring of fluid mud in ports and waterways using ultrasonic measurements in transmission and reflection geometry, including measurements with Distributed Acoustic Sensing (DAS), and using temperature measurements with Distributed Temperature Sensing (DTS). We performed the measurements in a laboratory on samples from the Port of Rotterdam, Port of Hamburg, and two synthetic clays.

Using ultrasonic transmission measurements with transducers directly inside fluid mud, we investigated the changes in the velocities of longitudinal (P-) and transverse (S-) waves and their possible relation to the yield stress during the consolidation. We observed no detectable change of the P-wave velocities during the consolidation of the fluid mud. We observed that the S-wave velocities exhibited a relatively strong increase after the fluid mud settles for a certain amount of time, in our study after 3 days. Comparing the estimated S-wave velocities to the concurrently estimated fluidic yield stress, we showed a positive correlation between the two. Our findings verify that the S-wave velocities increase with increasing yield stress caused by the fluid-mud consolidation and can thus be potentially used for indirect in-situ assessment of the yield stress.

Using ultrasonic reflection measurements with transducers, we investigated the direct estimation of the P- and S-wave velocities inside the fluid-mud layer. The source and receiver transducers were placed inside the water layer, but we showed that the kinematic influence of the water layer can be completely eliminated by retrieval of non-physical (ghost) reflections inside the fluid mud by application of seismic interferometry. Using the retrieved ghost reflections to estimate the layer-specific P- and S-waves velocities of the mud, we eliminated possible uncertainty due to salinity and temperature gradients of the water, which affect the velocity estimates using the usual seismic-reflection processing techniques. We show that the reflection-estimated velocities differ from the transmission-calculated values only by 1.4% and 0.3% for the P- and S-waves, respectively.

We also showed that DAS and DTS can be very effective in estimating the depth of the water/mud interface. We showed that a standard communication fiber is sufficient to achieve an accuracy in the estimated depth of the water/mud interface of 1.2 cm. This accuracy, to the best of our knowledge, is higher than what is achievable with any the currently used non-intrusive methods. Furthermore, we showed that the strength of the signal recorded with DAS is linked to changes in the shear strength of clays.

Acknowledgements

The research of X.M. is supported by the Division for Earth and Life Sciences (ALW) with financial aid from the Netherlands Organization for Scientific Research (NWO) with grant no. ALWTW.2016.029. The research of M.B. is supported by the Port of Rotterdam, Hamburg Port Authority, Rijkswaterstaat and SmartPort. The project is carried out also within the framework of the MUDNET academic network https://www.tudelft.nl/mudnet/.

Conflict of interest

The authors declare no conflict of interest.

Author details

Deyan Draganov[1*], Xu Ma[1], Menno Buisman[1,2], Tjeerd Kiers[1], Karel Heller[1] and Alex Kirichek[1,3]

1 Faculty of Civil Engineering and Geosciences, Delft University of Technology, Delft, The Netherlands

2 Port of Rotterdam, Rotterdam, The Netherlands

3 Deltares, Delft, The Netherlands

*Address all correspondence to: d.s.draganov@tudelft.nl

IntechOpen

References

[1] McAnally WH, Friedrichs C, Hamilton D, Hayter E, Shrestha P, Rodriguez H, Sheremet A, Teeter A, ASCE task committee on Management of Fluid mud. Management of fluid mud in estuaries, bays, and lakes. I: Present state of understanding on character and behavior. Journal of Hydraulic Engineering. 2007 Jan;133(1):9-22.

[2] Harbour Approach Channels Design Guidelines. In: Report of Marcom Working Group. 2014. p. 49.

[3] Kirichek A, Chassagne C, Winterwerp H, Vellinga T. How navigable are fluid mud layers. Terra et Aqua: International Journal on Public Works, Ports and Waterways Developments. 2018;151.

[4] Delefortrie G, Vantorre M, Eloot K. Modelling navigation in muddy areas through captive model tests. Journal of marine science and technology. 2005 Dec 1;10(4):188-202.

[5] Vantorre M. Ship behaviour and control in muddy areas: state of the art. InProceedings of the 3rd International Conference on Manoeuvring and Control of Marine Craft (MCMC'94), edited by GN Roberts and MMA Pourzanjani, Southampton 1994 Sep (pp. 7-9).

[6] Claeys S, De Schutter J, Vantorre M, Van Hoestberghe T. Rheology as a survey tool: We are not there yet. Hydro International. 2011;15(3):14-19.

[7] Kirichek A, Rutgers R. Monitoring of settling and consolidation of mud after water injection dredging in the Calandkanaal. Terra et Aqua. 2020; 160: 16-26

[8] Kirichek A, Shakeel A, Chassagne C. Using in situ density and strength measurements for sediment maintenance in ports and waterways.

Journal of Soils and Sediments. 2020 Feb 19:1-7.

[9] Hamilton EL, Bachman RT. Sound velocity and related properties of marine sediments. The Journal of the Acoustical Society of America. 1982 Dec;72(6):1891-1904.

[10] Schrottke K, Becker M, Bartholomä A, Flemming BW, Hebbeln D. Fluid mud dynamics in the Weser estuary turbidity zone tracked by high-resolution side-scan sonar and parametric sub-bottom profiler. Geo-Marine Letters. 2006 Sep 1;26(3): 185-198.

[11] Gratiot N, Mory M, Auchere D. An acoustic Doppler velocimeter (ADV) for the characterisation of turbulence in concentrated fluid mud. Continental Shelf Research. 2000 Sep 10;20(12–13): 1551-1567.

[12] Meissner R, Rabbel W, Theilen F. The relevance of shear waves for structural subsurface investigations. InShear waves in marine sediments 1991 (pp. 41-49). Springer, Dordrecht.

[13] Shapiro NM, Campillo M. Emergence of broadband Rayleigh waves from correlations of the ambient seismic noise. Geophysical Research Letters. 2004 Apr 16;31(7).

[14] Wapenaar K, Fokkema J. Green's function representations for seismic interferometry. Geophysics. 2006 Jul;71 (4):SI33-SI46.

[15] Draganov D, Campman X, Thorbecke J, Verdel A, Wapenaar K. Reflection images from ambient seismic noise. Geophysics. 2009 Sep;74(5):A63-A67.

[16] Draganov D, Ghose R, Ruigrok E, Thorbecke J, Wapenaar K. Seismic

interferometry, intrinsic losses and Q-estimation. Geophysical Prospecting. 2010 Mar 26;58(3):361-373.

[17] Draganov D, Heller K, Ghose R. Monitoring CO2 storage using ghost reflections retrieved from seismic interferometry. International Journal of Greenhouse Gas Control. 2012 Nov 1;11: S35-S46.

[18] King S, Curtis A. Suppressing nonphysical reflections in Green's function estimates using source-receiver interferometrySuppressing nonphysical reflections. Geophysics. 2012 Jan 1;77 (1):Q15-Q25.

[19] Breitzke M. Acoustic and elastic characterization of marine sediments by analysis, modeling, and inversion of ultrasonic P wave transmission seismograms. Journal of Geophysical Research: Solid Earth. 2000 Sep 10;105 (B9):21411-21430.

[20] Leurer KC. Compressional-and shear-wave velocities and attenuation in deep-sea sediment during laboratory compaction. The Journal of the Acoustical Society of America. 2004 Oct;116(4):2023-2030.

[21] Ballard MS, Lee KM, Muir TG. Laboratory P-and S-wave measurements of a reconstituted muddy sediment with comparison to card-house theory. The Journal of the Acoustical Society of America. 2014 Dec;136(6):2941-2946.

[22] Ballard MS, Lee KM. Examining the effects of microstructure on geoacoustic parameters in fine-grained sediments. The Journal of the Acoustical Society of America. 2016 Sep 8;140(3):1548-1557.

[23] Collins JA, Sutton GH, Ewing JI. Shear-wave velocity structure of shallow-water sediments in the East China Sea. The Journal of the Acoustical Society of America. 1996 Dec;100(6): 3646-3654.

[24] Ajo-Franklin JB, Dou S, Lindsey NJ, Monga I, Tracy C, Robertson M, Rodriguez Tribaldos V, Ulrich C, Freifeld B, Daley T, Li X. Distributed acoustic sensing using dark fiber for near-surface characterization and broadband seismic event detection. Sientific Reports. 2019; 9:1328.

[25] Jousset P, Reinsch T, Ryberg T, Blanck H, Clarke A, Aghayev R, Hersir GP, Henninges J, Weber M, Krawczyk CM. Dynamic strain determination using fibre-optic cables allows imaging of seismological and structural features. Nature Communications. 2018; 9: 2509.

[26] Lindsey NJ, Martin ER, Dreger DS, Freifeld B, Cole S, James RS, Biondi BL, Ajo-Franklin JB. Fiber-optic network observations of earthquake Wavefields. Geophysical Research Letters. 2017 Dec 16; 44(23):11792-11799.

[27] Wang HF, Zeng X, Miller DE, Fratta D, Feigl KL, Thurber CH, Mellors RJ. Ground motion response to an ML 4.3 earthquake using co-located distributed acoustic sensing and seismometer arrays. Geophysical Journal International. 2018 Jun;213(3):220-236.

[28] Yu C, Zhan Z, Lindsey NJ, Ajo-Franklin JB, Robertson M. The potential of DAS in Teleseismic studies: Insights from the goldstone experiment. Geophysical Research Letters. 2019 Feb 16; 46(3):1320-1328.

[29] Daley TM, Miller DE, Dodds K, Cook P, Freifeld BM. Field testing of modular borehole monitoring with simultaneous distributed acoustic sensing and geophone vertical seismic profiles at Citronelle, Alabama. Geophysical Prospecting. 2016; 64(5): 1318-1334.

[30] Mateeva A, Lopez J, Potters H, Mestayer J, Cox B, Kiyashchenko D, Wills P, Grandi S, Hornman K, Kuvshinov B, Berlang W, Yang Z,

Detomo R. Distributed acoustic sensing for reservoir monitoring with vertical seismic profiling. Geophysical Prospecting. 2014; 62(4):679–692.

[31] Zeng X, Lancelle C, Thurber C, Fratta D, Wang H, Lord N, Chalari A, Clarke A. Properties of noise cross-correlation functions obtained from a distributed acoustic sensing Array at Garner Valley, California. Bulletin of the Seismological Society of America. 2017 Jan 31; 107(2):603-610.

[32] Ravet F, Rochat E, Niklès M. BOTDA-based DTS robustness demonstration for subsea structure monitoring applications. Proc. SPIE 9634, 24th International Conference on Optical Fibre Sensors. 2015 Sep 28; 96345Z.

[33] Stork AL, Chalari A, Durucan S, Korre A, Nikolov S. Fibre-optic monitoring for high-temperature carbon capture, utilization and storage (CCUS) projects at geothermal energy sites. First Break. 2020 Oct; 38(10):61-67.

[34] Shao M, Qiao X, Zhao X, Zhang Y, Fu H. Liquid level sensor using fiber Bragg grating assisted by multimode fiber core. IEEE Sensors Journal. 2016 Jan 7;16(8):2374-2379.

[35] Xu W, Wang J, Zhao J, Zhang C, Shi J, Yang X, Yao J. Reflective liquid level sensor based on parallel connection of cascaded FBG and SNCS structure. IEEE Sensors Journal. 2016 Nov 16;17 (5):1347-1352.

[36] Yang C, Chen S, Yang G. Fiber optical liquid level sensor under cryogenic environment. Sensors and Actuators A: Physical. 2001 Oct 31;94 (1–2):69-75.

[37] Wei M, McGuire JJ, Richardson E. A slow slip event in the south Central Alaska subduction zone and related seismicity anomaly. Geophysical Research Letters. 2012 Aug 16;39(15).

[38] Shizhuo Y, Ruffin PB, Francis TS. Fiber optic sensors. Talor & Francis Group. 2008;479.

[39] Shakeel A, Kirichek A, Chassagne C. Rheological analysis of mud from port of Hamburg, Germany. Journal of Soils and Sediments. 2020; 1-10.

[40] Shakeel A, Kirichek A, Chassagne C. Yield stress measurements of mud sediments using different rheological methods and geometries: An evidence of two-step yielding. Marine Geology. 2020; 106247.

[41] Draganov D, Hunziker J, Heller K, Gutkowski K, Marte F, high-resolution ultrasonic imaging of artworks with seismic interferometry for their conservation and restoration. Studies in Conservation. 2018; 63(5):277-291.

[42] Lindsey NJ, Rademacher H, Ajo-Franklin JB. On the broadband instrument response of fiber-optic DAS arrays. Journal of Geophysical Research: Solid Earth. 2020 Feb;125(2): e2019JB018145.

[43] Li W, Bao X, Li Y, Chen L. Differential pulse-width pair BOTDA for high spatial resolution sensing. Optics express. 2008 Dec 22;16(26):21616-21625.

[44] Kuvshinov B. Interaction of helically wound fibre-optic cables with plane seismic waves. Geophysical Prospecting. 2016; 64(3): 671–688.

[45] Ayres A, Theilen F. Relationship between P- and S-wave velocities and geological properties of near-surface sediments of the continental slope of the Barents Sea. Geophysical prospecting. 2001 Dec 24;47(4):431-441.

[46] Snieder R. Extracting the Green's function from the correlation of coda waves: A derivation based on stationary phase. Physical review E. 2004; 69(4): 046610.

Chapter 8

Activated Flooded Jets and Immiscible Layer Technology Help to Remove and Prevent the Formation of Bottom Sediments in the Oil Storage Tanks

Georgii V. Nesyn

Abstract

Two flooded jet methods of tank bottom sediments caving based on either screw propeller generation or nozzle jets generated with entering crude head oppose each other. The comparison is not advantageous for the first one. Exceptionally if crude oil contains some concentration of high molecular weight polymer which can perform Drag Reduction. In this case, the jet range increases by many times, thus, upgrading the capability of caving system. Preventing the sedimentation of crude oil heavy components may be put into practice with Immiscible Layer Technology. Before filling the tank with crude oil, some quantity of heavy liquid, that is immiscible with all the components of crude oil, is poured into the tank. The most suitable/fit for purpose and available liquid is glycerin. Neither paraffin and resins, nor asphaltenes can penetrate through the glycerin layer to settle down at the tank bottom because of its density, which is equal to 1.26 g/cm^3. Instead, sediments are concentrated at/on the glycerin surface and when it is heated in external heat exchanger all the sediments ought to move upwards with the convection streams. Thus, no deteriorate sediment is formed in the tank bottom.

Keywords: crude oil, storage tank, residue, sludge, sediments, crude heavy components, flooded jet, washing out, drag reducing agents, immiscible heavy liquid, glycerin, external heat exchanger, convection streams

1. Introduction

Long-term crude oil storage in the tanks, as well as transportation by tank wagons and tankers, is accompanied by agglomeration and precipitation of water, inorganic impurities, paraffin, resins and asphaltenes. Organic part of the sediment is a product of crude oxidation that occurs in the course of oil contact with air and is followed by condensation reactions. Inorganics and water are caught into oil during production and transportation. Temperature drop in the tank makes some of the heavy components of crude oil insoluble. They gravitate onto the bottom in aggregate with inorganics and water. Bottom sediment reduces effective tank volume, promotes corrosive action, complicates oil transfer and water drain out. As heavy crude fraction is growing step by step in the whole production volume, the problem of bottom sludge in the oil storage tank will enhance.

2. The bottom residue washing out technologies

Mechanical methods of sludge disposal usually include scraping and auger techniques. In complicated cases, a technological pass-through is made for a robot-machine [1]. The personnel in this case must work using supplied-air respirators. As mechanical cleaning is accompanied by a number of problems, such as pretreat-ment, degassing, manual labor in gas hazard conditions, oil sludge disposal, all of which making it rather expensive and time-consuming, it is only applied in the case of maintenance shutdown or tank repurposing.

Now Transneft Company makes extensive use of screw propellers, both for caving bottom deposits and prevention of their accumulations. They are fixed in a manhole of a tank first circle and make a flooded jet of oscillating direction within the sector of about 60° in a plane parallel to the tank bottom [2]. However, this method has downsides, the biggest of which is quite large energy consumption that makes the use of "Diogen" screw propeller for prevention of bottom sediments formation not feasible. Other disadvantages [3]: peripheral propeller location, i.e. at the tank wall, which reduces the efficiency of the impact on the bottom area close to the opposite wall, as well as the occurrence of vibrations and increased stress in the welded joint of the inlet-distribution nozzle. In addition, in the case of high-viscos-ity oil, mixing can be difficult due to reduced mixing efficiency. Purely mechani-cal cleaning methods are complemented, if necessary, by physical and chemical methods, which allow for more comprehensive disposal of heavy oil components. Among such methods are the action of solvents, reagents and heated oil [4]; warm water containing surfactants [5], etc. As a result of this action, deposits are heated, liquefied and homogenized with subsequent separation of hydrocarbons by means of filter presses [6], centrifuges [7], and other methods. However, almost always after hydrocarbons are removed, solid residues are buried, which has an adverse impact on the environment. The cleaning procedure involves emptying and decom-missioning of the tank for a period of two to seven weeks depending on the tank volume, amount of sediments and season of the year.

3. Prevention of bottom sediments' formation

As a rule, it is more beneficial to prevent the formation of sediments rather than to clean the already settled ones. For these purposes, for instance, a tank can be equipped with either jet hydraulic mixers (**Figure 1**) or a washout unit where the inlet nozzles fit with washout heads that generate flooded crude oil jets directed along the tank bottom. One of the options for washout heads is a plate lying on the tank bottom that, when being subjected to crude oil pressure, elevates, thus, forming a fan-shape jet.

Both washout systems are characterized by cost efficiency due to the use of the part of potential flow energy in the pipeline section preceding the tank inlet and quite high hydrodynamic capacity exceeding the capacity of propeller type mixers. Let us focus on washout heads system, the efficiency of which, as shown in paper [8], can be significantly enhanced without additional capital investments and which can be used not only to prevent tank bottom sediments formation but also for tank cleaning.

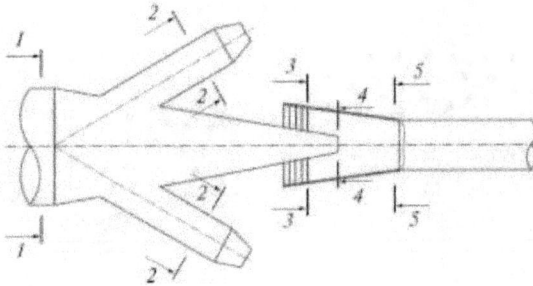

Figure 1.
Diagram of cross sections of mixer for calculations: 1 – Supply nozzle; 2 – Lateral nozzles; 3 – Confusor;
4 – Central nozzle; 5 – Inlet to the mixing chamber.

Under standard operational conditions, areas covered by washout heads should overlap. However, in the case of low oil pressure or if a head gap gets clogged, the areas of coverage of fan-shaped jets do not overlap and stagnation zones are formed where a build-up of bottom sediments occurs. In our case, for a number of reasons, sediments occupied about 20% of the useful volume of the tank. And activation of flooded jets of oil, coming out of washout heads, was attempted to be made through injection of high molecular weight oil soluble polymers into the pipeline section preceding the tank inlet [8].

It is known that polymer additives capable to reduce the hydrodynamic resistance of liquids, when introduced into the flow, significantly increase the compact part of flooded and free jets. Increase in the range of open jets is used, for example, in firefighting and water jet rock destruction. In the latter case, the aqueous polymer solutions perform similarly to sand slurries that have abrasion wear effect on perpendicular plate [9] due to polymeric associates present in the solution [10]. The size of such supermolecular features, according to the authors, is about 1 mm, and their relaxation time is about $1 \cdot 10^{-3}$-$1 \cdot 10^{-2}$ s. Associates behave like drops of ordinary fluid in deformation processes with characteristic time exceeding the aforementioned time. If the deceleration time is much shorter than their relaxation time, associates behave like solid particles. Analysis of thin sections of washout zones in metal plates treated with high-speed jets of aqueous polymer solutions indicate in favor of the impact nature of their destruction.

Something similar can occur in the flooded oil jets upon introduction of linear polymers molecules of high molecular weight into the flow: length of the compact part of a stream considerably increases, thus, the action area of a washout head increases by several times. On the other hand, in a compact part of jet, macromolecules and their associates, being guided by the flow, can form anisotropic structures of "thread" or "needle" types. Such inclusions, when encountering a perpendicular obstacle, at sharp braking may act as solid particles of an elongated shape that "dig" the sediment layer (**Figure 2**).

An additive of high-molecular octene-decenoic copolymer earlier used in experiments to increase the capacity of oil pipelines, was used as an activator for flooded jets. The copolymer solution in naphta was applied, because it takes a short time to get a mixture with oil flow. Sediments were washed out according to the scheme shown in **Figure 3** (here 1 - tank; 2 - washout heads; 3 - container with polymer solution; 4 – dosing pump; 5 - extraction pump).

Check gauging of the level of sediments made at five points - at the metering manhole, three inspection and one central manhole - showed that the average level in the first tank made 195 cm, and in the second tank - 206 cm, while the whole

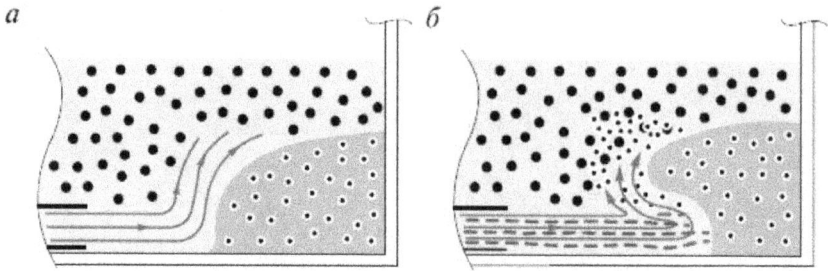

Figure 2.
Presumable picture of the flooded jet action without the polymer (a) and in the presence of the polymer (6).

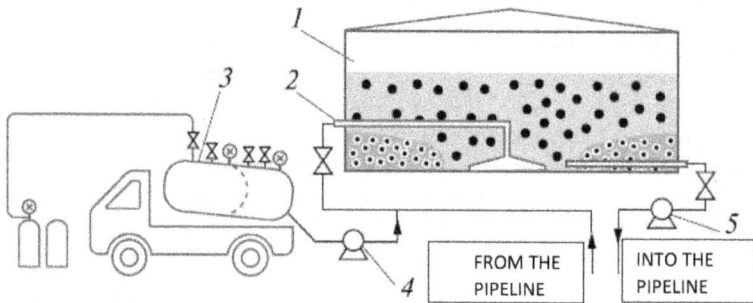

Figure 3.
Scheme of the bottom sediments washout by polymeric activators of flooded jets.

height of the tank was 1000 cm. The washout was carried out in a flow regime, i.e., when oil was simultaneously pumped into and out of the tank. The oil rate was about 2000 m³/hour. The level of oil above the sediment level was kept within the range of 1 to 1.5 m. The control 12 hour washout with clean oil performed through the washout heads gave practically no changes either in the height of the sediment level or in the oil composition at the inlet and outlet of the tanks.

The polymer solution was dosed into the receiving pipeline at a distance of 200 m from the tank, the polymer concentration in oil was about 50 ppm (parts per million). The beginning of dosing was accompanied by the growth of water presence in oil at the outlet of the tank: at an average concentration of 0.18% at the outlet, it exceeded 2%. It gave evidence of the dispersion of bottom sediments that contained water. In 10 hours the dosing of the concentrate was stopped, but the circulation regime was maintained for another 10 hours to complete the hydraulic transport of washed out sediments outside the tank. The tank was then filled with oil and allowed its contents to settle for 24 hours. The check measurements made after that showed that the sludge level after washing out made 80 cm. About half of the concentrate had been used.

In order to reach maximum efficiency, the second half of the concentrate was used to wash out the second tank. The cycle of works was repeated in the same order. The water content at the outlet reached 2.7%. Check measurements of sediment level made after cleaning showed an average value of 69 cm. Thus, as a result of the experiment, about 4000 tons of bottom sediments were removed and mixed with oil from two 20,000 ton vertical tanks. The cleaning procedure including preparation activities took ten calendar days. The interaction of the activated flooded jet with the sediments was so intense that the external tank wall warmed up to a temperature of about 40°C within the height of sediments location. Dissipation of mechanical energy of the "reinforced" jet into the heat energy, apparently, causes autoacceleration of the washout process due to softening the object under the impact of a flooded jet.

The following advantages of the considered cleaning method should be noted:

- useful settlement components are preserved;

- no soil contamination;

- simplicity of hardware design/equipment required;

- fire safety, as there are no works performed inside the tank;

- the tank is taken out of operation for a short time: the cleaning procedure itself takes about one week;

- no personnel are exposed to harmful fumes during the cleaning process.

The method of activation of oil flooded jets can be used both for tank cleaning and prevention of bottom sediments accumulation.

4. Immiscible layer

The idea of an immiscible layer is a separate topic, as it offers a fundamentally new approach to the prevention of bottom sedimentation [11]. It is suggested to use a layer of a heavy liquid covering the bottom of the tank and preventing the settling of heavy oil and water components.

Immiscible layer fluid should meet three requirements: it should be immiscible to oil hydrocarbons; it should have a greater density than oil and its boiling point should be high enough. Polyatomic alcohols are most suitable for this role. Of these, glycerin (**Figure 4**) is the most applicable for it is available and has a density of 1.26 g/cm^3. Asphaltenes that is the heaviest component of crude oil has a density not more than 1.1 g/cm^3.

The liquid circulates through an external heat exchanger that maintains the immiscible pad temperature sufficient for thermal convection of the lower oil layers (**Figure 5**). In this case, heavy asphaltenes will not accumulate on the glycerin surface, and, due to thermal convection, will be distributed along the bulk.

The water proposed for the role of such a liquid [12] is not quite suitable due to corrosion activity and low density: heavy resins and asphaltenes with a density of about 1.1 g/cm3 will penetrate through the water layer and settle on the bottom.

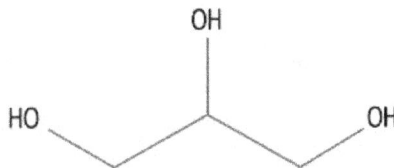

Figure 4.
Structural formula of glycerin.

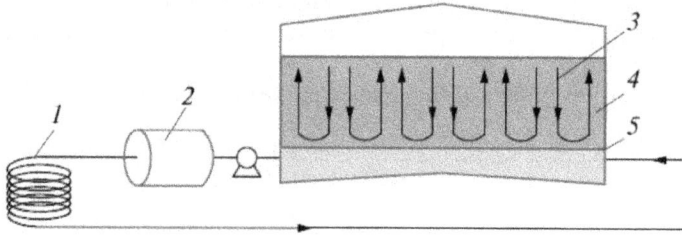

Figure 5.
Diagram of oil tank with immiscible layer.

Glycerin is not mixed with oil hydrocarbons and has a density of 1.26 g/cm3. Water and inorganic impurities enter the glycerin layer. Dissolved moisture is periodically removed from the glycerin by an external drying device (not shown in **Figure 5**). As far as mechanical impurities are concerned, they will be retained within the viscous glycerin for some time, and provided quite intense circulation, most of them will be collected at the external filter. Glycerin does not cause corrosion and its high density will protect the bottom of the tank from asphaltenes and resins settling.

The bottoms of oil tanks with the capacity of more than 5 thousand tons have, as a rule, a conical shape with a slope from the center of 1:100. The height of the immiscible layer should exceed the height of the bottom cone in the preferred case by at least 10 cm, so that the whole surface is covered with heavy immiscible layer fluid (see **Figure 5**). Based on this, the minimum glycerin volume for a 20 thousand tons tank should be 292 m^3, for a tank of 50 thousand tons - 875 m^3. The maximum volume of the hydraulic cushion is calculated from the economic feasibility: on the one hand, the formation of sediments and corrosion of the tank bottom are prevented, on the other hand, the layer volume is the "dead volume" of the tank cut off from the commodity transactions. The level of the distribution nozzle should also ensure that glycerin does not get into the pumped out oil.

All manipulations with the immiscible layer liquid can be carried out at external devices without tank shutting down from operation, and the complete set of the equipment on glycerin heating and cleaning can be used for handling of several tanks.

Glycerin is non-toxic and dissolves water together with salts. Moreover, its cost is not high, as it is a waste of biodiesel production from vegetable oils. Currently, a kilo of glycerin costs about $1 and there is a downward trend in the cost.

This method of preventing the formation of bottom sediments can multiply the period between tank cleanups and the lifetime of the tanks. Accordingly, losses of heavy oil fractions will reduce, and operation of the tank farm will be more efficient and environmentally safe.

5. Conclusion

Bringing into development of bituminous oil fields leads to "weighting" of oil pumped through the main oil pipelines. High viscosity oil has an even greater tendency to form sediments during storage. The use of heated tanks partially solves the problem, but is associated with high energy consumption. Forced circulation screw devices can be ineffective in a highly viscous medium. In this regard, the research described in the article may be useful for solving the problems that arise.

Author details

Georgii V. Nesyn
Transneft R&D, LLC, Russia

*Address all correspondence to: nesyngv@niitnn.transneft.ru

IntechOpen

References

[1] Gimaletdinov G.M. & Sattarova D.M., 2006: Methods for cleaning and preventing accumulation of bottom sediments in the tanks. Neftegazovoye delo = Oil and Gas Business. 2006. pp. 1-12. Available at: http://ogbus.ru/authors/Gimaletdinov/ Gimaletdinov_1.pdf (accessed: November 2, 2020). (In Russ.).

[2] Valiyev M.R., 2014: Up-to-date methods for cleaning the cavity of vertical steel tanks from bottom sediments. Available at: www.lib.tpu. ru/fulltext/c/2014/C11/V2/236.pdf (accessed: November 2, 2020). (In Russ.).

[3] Galiakbarova E.V. et al., 2015: Galiakbarova E.V., Bakhtizin R.N., Nadrshin A.S., Galiakbarov V.F. Safe and energy-efficient elimination of sedimentation during storage of oil in the tanks. Neftegazovoye delo = Oil and Gas Business. 2015. Vol. 13. № 4. pp. 142-148. (In Russ.).

[4] Isyanov F.T. et al., 2010: Isyanov F.T., Korkh L.M., Tarraf A., Rasvetalov V.A. Patent RF. № 2442632. The method for cleaning tanks from oil sludge. Applied: March 12, 2010. Published: February 20, 2012. Bulletin № 5. (In Russ.).

[5] Goss M. L., 1992: Patent US. № 5085710. The method of using aqueous solutions of chemical agents for the recovery of hydrocarbons and minimization of losses in oil tanks. Applied: October 31, 1989. Published: February 4, 1992.

[6] Bakhonina E.I., 2015: Up-to-date technologies of processing and recovery of hydrocarbon-containing waste. Information 2. Physico-chemical, chemical, biological methods of utilizationand neutralization of hydrocarbon-containing waste. Bashkirskiy khimicheskiy zhurnal = Bashkiria Chemical Journal. 2015. № 2. pp. 41-49. (In Russ.).

[7] Davis G.B. et al.,1993: Davis G.B., Goss M.L., Schoemann P., Tyler S.S. Crude oil tank cleaning process recovers oil, reduces hazardous wastes. Oil and Gas Journal. 1993. 13/XIL. Vol. 91. № 50. pp. 35-39.

[8] Nesyn G.V., 2007: Development of high-molecular additives that increase the throughput of oil pipelines: Thesis. ... Doctor of Chemical Sciences. Kazan: Kazanskiy gosudarstvennyy tekhnologicheskiy universitet, 2007. (In Russ.).

[9] Shtertser A.A. & Grinberg B.E., 2013: Effect of the hydroabrasive jet on the material: hydroabrasive wear. Prikladnaya mekhanika i teoreticheskaya fizika = Applied Mechanics and Theoretical Physics. 2013. Vol. 54. № 3. pp. 191-201. (In Russ.).

[10] Kudin A.M. et al., 1973: Kudin A.M., Barenblatt G.I.,Kalashnikov V.N., Vlasov S.A., Belokon' V.S. Nature Physical Science 245 (145) 95-96 (1973)

[11] Nesyn G.V., 2017: Patent RF. № 2637915. Method for preventing the formation of bottom sediments in the tanks for oil storage and / or transportation. Applied: April 18, 2016. Published: December 7, 2017. Bulletin № 34. (In Russ.).

[12] Gushchin V.V., et al., 2006: Gushchin V.V., Yakovenko G.V., KasharabaO.V.,OrlovG.I.,KoshcheyevV.I., Berlin M.A., Grabovskiy Yu.P. Patent RF. № 2286297. Method of oil storage and the means for its implementation. Applied: May 5, 2005. Published: October 27, 2006. (In Russ.).

Chapter 9

Study of Water and Sediment Quality in the Bay of Dakhla, Morocco: Physico-Chemical Quality and Metallic Contamination

Mimouna Anhichem and Samir Benbrahim

Abstract

The present study contributes to the evaluation of the impact of the various activities developed around the Bay of Dakhla in Morocco through the study of the physico-chemical quality of the waters and sediments of the Bay. For this purpose, a spatial and temporal monitoring of the physicochemical and metallic pollution indicator parameters was conducted between May 2014 and March 2015. The main physicochemical descriptors of water quality were monitored, namely: temperature, salinity, pH, dissolved O2, nutrients (ammonium, nitrites, nitrates, phosphates) and chlorophyll (a). A qualification of the waters of the Bay was drawn up based on water quality assessment grids. The quality of the sediments was assessed through the determination of granulometry, the total organic carbon content and the contents of the main metallic trace elements (cadmium, lead, mercury, chromium, copper and zinc). The results of the present study show the beginning of nutrient enrichment of the water bodies of the bay, especially the stations located near the urban area, where 1.83 mg l^{-1} of nitrates, 0.37 mg l^{-1} of phosphate and 7.42 μg l^{-1} of chlorophyll (a) were recorded. For the sediment, the maximum concentrations of metallic trace elements were recorded in the station near the harbour basin. These results allowed to establish a quality grid for the waters of the bay, generally qualified as "Good", except for the sites located near the urban area for which the quality is qualified as "Average". The sediment quality of the bay was assessed according to the criteria established by the Canadian Council of Ministers of the Environment. The levels of metallic trace elements remain below the toxicity thresholds, except for the sediments taken from the harbour basin.

Keywords: Dakhla Bay, sediments, contaminations, hydrological parameters, trace metals, toxicity

1. Introduction

Paralic environments constitute a transition space between continental and marine ecosystems. They are areas of exchange and transfer of energy and nutrients, very favorable to the development of biological abundance. These environments are therefore the most important marine areas, but also the most vulnerable areas. They present an extremely complex dynamic, influenced by the open ocean and the terrestrial environment. These environments are bodies of water that are often confined, poorly renewed and therefore naturally vulnerable and whose balance can be rapidly modified under the influence of natural or anthropogenic factors [1, 2]. The physico-chemical properties of water masses, as well as the phenomena of tides, swell, and various types of currents, modify the nature of the fauna and flora [3]. The preservation of these fragile environments, of major socio-economic interest, therefore requires knowledge of the processes controlling their evolution [4, 5].

Monitoring the physico-chemical parameters of water bodies and assessing the levels of sediment contamination can help reduce the constraints imposed on these ecosystems and predict possible scenarios to preserve these environments.

For the water bodies the parameters temperature, salinity, pH as well as dissolved oxygen condition the presence of the species according to their preference. Depending on the degree of disturbance of these parameters, the variations can have an influence on the movement of species (e.g. barrier to migration) or have more permanent impacts by disturbing the physiological evolution of organisms (e.g. problems of growth, reproduction, ...) [6–8]. In addition, nutrients and chlorophyll (a), which are hydrological descriptors essential for the study or characterization of the water masses of a marine ecosystem, can have repercussions on human activities such as fishing and shellfish farming because their availability conditions the primary production on the basis of which the whole biological activity of the environment then develops [7].

The physico-chemistry of water is considered to be a supporting element for biology, i.e. the quality thresholds to be determined must transcribe the environmental conditions that allow or not the different biological compartments to be in good condition. Thus, it is necessary to know the requirements of the living in terms of temperature, salinity, dissolved oxygen and nutrients, which implies an analysis of the links between physico-chemical parameters and biology as well as their preferences. Threshold grids of a few physico-chemical parameters and a classification from "very good" to "poor condition" in relation to the needs and tolerances of the ichthyofauna have been defined by the European Water Framework Directive [9].

The nutrient indicator proposed by [10] includes only dissolved inorganic nitrogen (DIN) concentrations, which includes ammonium, nitrates and nitrites. ICES [11].

For phosphate ions, the grid for assessing water quality and its suitability for the natural functions of aquatic environments is described by [12].

For sediments, they are the memory of hydro-sedimentary events and constitute both a place of accumulation and emission of pollutants. Any change in the quantities or nature of inputs (terrigenous, industrial and urban) to the environment is recorded in sediments [13]. The Canadian Council of Ministers of the Environment has established two reference values for trace metal elements in marine sediments. These reference values are defined by an Effect Threshold Concentration (ETC) and a Probable Effect Concentration (PEC). These two reference values have been retained among the new sediment quality criteria, but are not sufficient to determine all the thresholds necessary for sediment management. Three other quality criteria were therefore defined later, namely the Rare Effect Concentration (REC), the Occasional Effect Concentration (OEC) and the Frequent Effect Concentration (FEC). Together, these criteria provide a screening tool to assess the degree of sediment contamination. These criteria can prevent the contamination of sites that are vulnerable to anthropogenic contaminant input [14].

The biological richness of the Moroccan coastline is linked to the presence of a large number of paralic zones, such as estuaries, lagoons and bays. The latter are coveted, despite their fragility requiring special management. Dakhla Bay is considered one of the most important in Morocco, both in terms of its surface area and its fish stocks. It represents an ecosystem with strong potential in terms of aquaculture, especially shellfish farming. Moreover, it is characterized by vast beaches and permanent winds, favorable to the development of water sports. This marine domain, so rich in fauna and flora, plays an important socio-economic role for Morocco and therefore imposes a commitment for its protection and the preservation of its resources for future generations. The bay is subject to numerous anthropic pressures, particularly since the opening of the port of Dakhla in 2001 and the extension of industrial activities related to fishing, aquaculture and tourism. These different activities invite questions as to their possible negative impacts on the ecosystem of the bay and its balance.

The objective of this study is to carry out a diagnosis of the state of "health" of Dakhla Bay. Contamination assessment has focused on water and sediment quality indicators. The main physicochemical descriptors of water quality, namely: temperature, salinity, pH, dissolved O2, nutrients (ammonium, nitrites, nitrates, phosphates) and chlorophyll (a) were monitored. The quality of the sediments was assessed through the determination of granulometry, total organic carbon content, the contents of the main metallic trace elements (cadmium, lead, mercury, chromium, copper and zinc). A qualification of the waters and sediments of the bay was drawn up based, successively, on the evaluation grids resulting from the European Water Framework Directive [9] and Marine Sediment Quality Criteria according to the Canadian Council of Ministers of the Environment [14].

2. Materials and methods

2.1 Study area

Dakhla Bay is located on the Atlantic coast of southern Morocco. 37 km long and about 13 km wide, the Bay is relatively narrow and open to the ocean to the south. Oriented NE–SW, it is bounded on the Atlantic Ocean side by the peninsula of Oued Ad Dahab, formed by sandy dunes (**Figure 1**). Dakhla Bay is classified as a Site of Biological and Ecological Interest (SBEI), an Area of International Importance for the Conservation of Birds (IBA) and a RAMSAR site (site recognized as a wetland of international importance) [15]. Unique in North Africa, it is both a migration relay and a wintering and nesting area for thousands of water-birds [16].

2.2 Sampling points and sampling frequency

Sampling sites were selected to cover the entire bay, particularly areas influenced by human activities (e.g. fishing, aquaculture, tourism, urban planning). For the water compartment, eleven stations were selected. Sampling was carried out at seasonal intervals between May 2014 and March 2015. A total of four sampling and measurement campaigns were carried out. For sediments, five stations were sampled during the winter of 2015 (**Figure 1**).

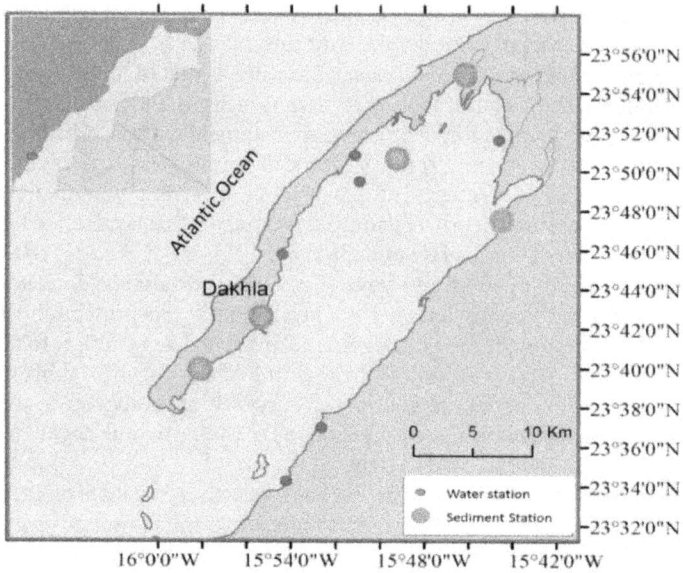

Figure 1.
Location of sampling stations in Dakhla Bay.

2.3 Sampling and analysis

2.3.1 Water compartment

- Physical parameters
  Temperature, salinity and pH were measured *in situ using* a portable multi-parameter probe WTW LF 197 (accuracy 0.1 units).

- Dissolved oxygen
  Dissolved oxygen was determined by Winkler's chemical method. Sampling was carried out in special glass vials with ground glass stoppers of known volume. Oxygen fixation is carried out on site by addition of the reagents. The method is designed to isolate the sample from air and fix the dissolved oxygen as quickly as possible by reaction with a precipitate of manganese hydroxide formed in the sample. Through a succession of reactions, an iodine solution is obtained which is easily and accurately quantified and has a concentration proportional to that of the oxygen initially present. The results are expressed in mg l^{-1} [7].

- Nutrient analysis
  Water samples for nutrient analysis (ammonium, nitrite, nitrate, phosphate) were taken sub-surface at a depth of about 0.5 m in clean polyethylene bottles previously rinsed with water to be analysed. The vials were then stored in a cooler at a temperature of approximately 4°C in the dark and transported to the laboratory for analysis. Nutrients were dosed by colorimetry according to the protocols described by [7].

- chlorophyll (a)
  For the determination of chlorophyll (a), one litre of water was filtered as soon as it arrived at the laboratory, under vacuum on a membrane (47 mm Whatman GF/C filter). The filters were then immersed in a solvent (90% acetone solution

10 ml volume) to dissolve the chlorophyll. The chlorophyll biomass (chlorophyll content (a) in µg l^{-1}) was estimated by spectrophotometry [7].

2.3.2 Sediment compartment

Sediment samples were taken from the surface layer using a hand corer. The contents of the corer were placed in a food-grade plastic bag, transported to the laboratory in coolers at approximately 4°C and then stored in the freezer at −20°C until analysis.

- Particle size analysis
  After drying at 40°C, the sediment samples were subjected to a conventional particle size analysis. A fraction of each sample was washed on 2 mm and 0.063 mm mesh sieves to separate the following three particle size classes [17]:

  ○ Class of "Rudites", with a particle diameter greater than 2 mm;

  ○ Class of "Arenites", with a particle diameter between 2 and 0.063 mm;

  ○ Class of "Lutites", with a particle diameter of less than 0.063 mm.

- Total organic carbon
  The determination of total organic carbon was carried out by indirect titration using the Walkley-Black method. This consists of oxidation of the organic carbon by a mixture of potassium dichromate and sulphuric acid. After the reaction, the concentration of total organic carbon is determined by measuring the excess dichromate. Titration is done by Mohr salt using diphenylamine as a colour indicator [18].

- Trace metals
  The extraction of metallic trace elements (cadmium, lead, mercury, chromium, copper and zinc) from the sediments was carried out by microwave mineralization using a mixture of strong acids: HNO3-HF and HCl [19]. The solutions obtained were analysed by Thermo iCAP Q Series Plasma Mass Spectrometry (ICP-MS). A certified reference material (IAEA-158) and a blank were analyzed with each mineralization series and are used for quality control and reliability of results.

2.4 Statistical processing of data

In order to highlight the relationships that can exist between the environmental factors studied (physico-chemical parameters and metal concentration) and the activities carried out at the bay, the obtained data were processed using XLSTAT 2016.06 software.

3. Results and discussions

3.1 Water compartment

3.1.1 Temperature

The surface water temperature values recorded *in situ* during this study allowed us to illustrate spatial and temporal variations in this parameter (**Figure 2**). The

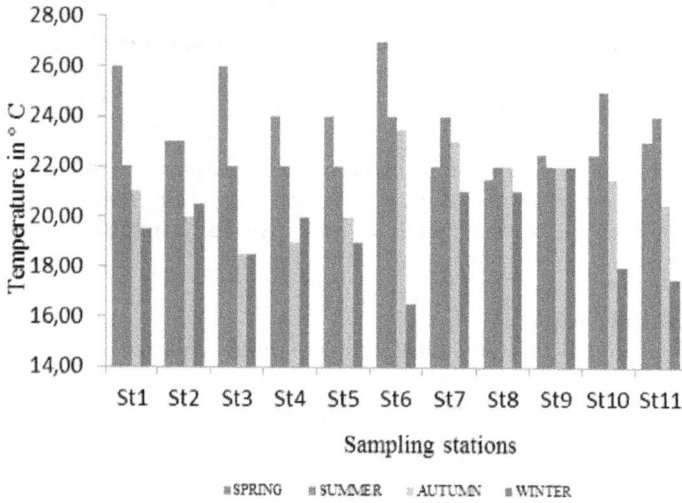

Figure 2.
Spatio-temporal evolution of the temperature at Dakhla Bay.

mean value recorded for the entire area was 21.8°C. The greatest thermal amplitude was observed at the Dunablanca station, with a minimum of 16.5°C recorded in winter 2015 and a maximum of 27.0°C recorded in spring 2014. This variation is quite normal given the shallow depth of the basin, which facilitates air-water exchanges.

3.1.2 Salinity

Salinity (**Figure 3**) ranges from 33.0 to 40.5 PSU, with a mean value of 36.9 PSU. The results obtained are in perfect agreement with previous work [20] in which a mean salinity of 36.9 PSU was recorded for the entire area, with an increasing gradient from the ocean to the bottom of the bay. The maximum value of 40.5

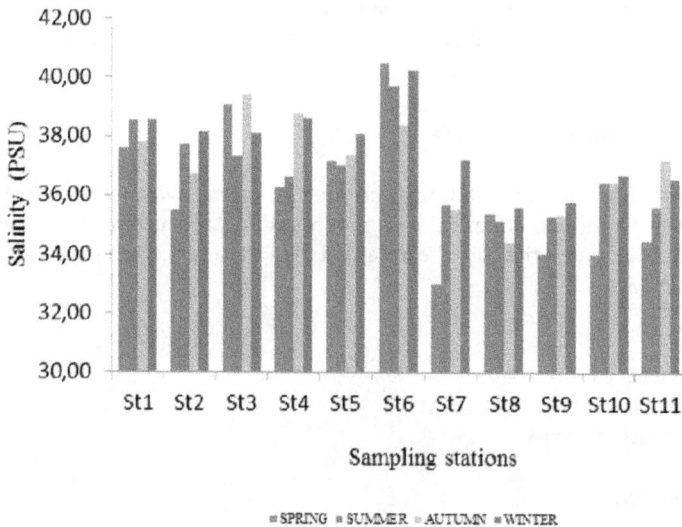

Figure 3.
Spatio-temporal evolution of salinity at Dakhla Bay.

PSU was recorded during spring 2014 at the Dunablanca site, where the highest temperature was recorded. This value would be explained by the location of the site at the bottom of the bay, characterized by slow renewal of marine waters [21] and a shallow depth favouring greater evaporation when the temperature increases. The value of 33.0 PSU was recorded during the spring of 2014 at the station in the urban area. Dilution of seawater by wastewater discharge would be the main source of the decrease in salinity at this sampling point.

3.1.3 pH

During the present study the values recorded for pH oscillate around 7.8 in the urban area near the discharges and 8.3 in the Lasargua station, with an average of 8.0 during the four seasons (**Figure 4**). These values are of the same order of magnitude as those reported in previous studies of the bay [22, 23].

3.1.4 Dissolved oxygen

The dissolved oxygen concentrations recorded range from 7.23 mg l^{-1} and 10.83 mg l^{-1}, with a mean value of 8.74 mg l^{-1} (**Figure 5**). This good oxygenation is mainly due, on the one hand, to the strong currents in the southern part of the bay where the strong exchanges with the ocean take place and, on the other hand, to the winds that stir the surface waters. The lowest levels were recorded at stations 7 and 8. However, oxygen levels remain above the required level despite the proximity of these urban discharge stations.

3.1.5 Ammonium

The maximum value recorded in the bay for the ammonium parameter is 0.31 mg l^{-1} and corresponds to the sample taken at the point in the urban area (city centre), and the minimum value is 0.00 mg l^{-1}, with an average of 0.06 mg l^{-1}. Ammonium is considered to be the hub of the nitrogen cycle in coastal ecosystems. Its concentrations in marine waters are often below 0,01 mg l^{-1} or even undetectable. Ammonium is mainly a tracer of urban and industrial discharges [24], the rise

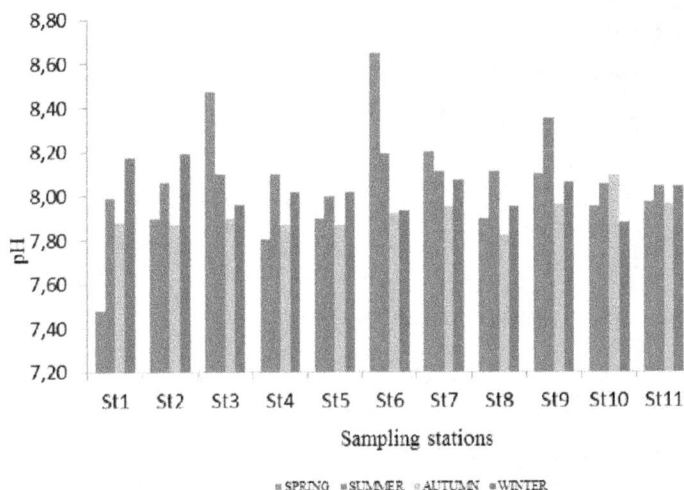

Figure 4.
Spatio-temporal evolution of the pH at Dakhla Bay.

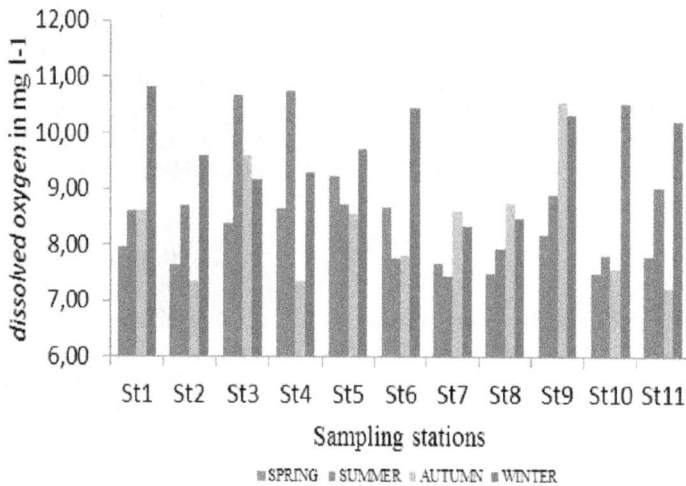

Figure 5.
Spatio-temporal evolution of dissolved oxygen at Dakhla Bay.

recorded during spring and summer (**Figure 6**) is mainly due to wastewater from the industrial area, port and some outfalls located in the urban area. On the other hand, the absence of ammonium in most stations during autumn and winter is related to the temperature drop that slows down biological activities and probably to the flow of urban and industrial discharges near the urban area. Another study on the character-ization of these discharges is being carried out to identify their impacts on the bay.

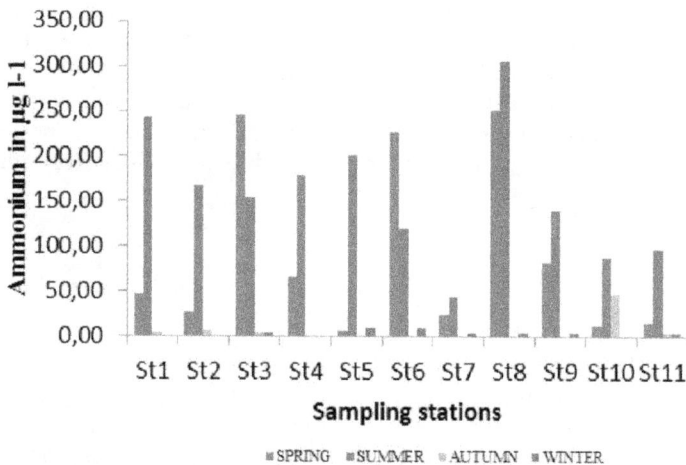

Figure 6.
Spatio-temporal evolution of ammonium at Dakhla Bay.

3.1.6 Nitrites

Nitrite concentrations in the study area range from a maximum of 0.04 mg l^{-1} to a minimum of 0.00 mg l^{-1}. The mean value of this parameter is 0.01 mg l^{-1} (**Figure 7**). The spatial distribution of nitrite is dependent on the proximity of the stations studied to sources of enrichment of the environment by this element.

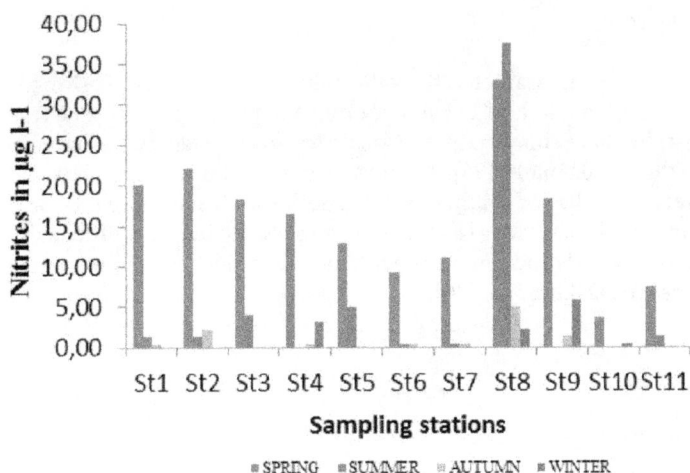

Figure 7.
Spatio-temporal evolution of nitrites at Dakhla Bay.

The dominance is observed during spring and summer at the point in the urban area (downtown) subject to urban and industrial discharges.

3.1.7 Nitrates

The values recorded for nitrates range from a maximum of 1.83 mg l^{-1} in the urban area, with a minimum of 0.00 mg l^{-1} recorded at most stations, especially during autumn and winter, with an average of 0.12 mg l^{-1} (**Figure 8**). The absence of nitrate ions at most sites in the bay during most of the year is only a strong signal of the significant biological activity within the bay. For the "urban area" station that recorded the maximum value, the increase in this element during most of the year means an enrichment of the environment by organic matter which, in the presence of nitrifying bacteria and oxygen, is transformed into nitrite and then into nitrate. This increase is certainly due to inputs from discharges that are close to the area.

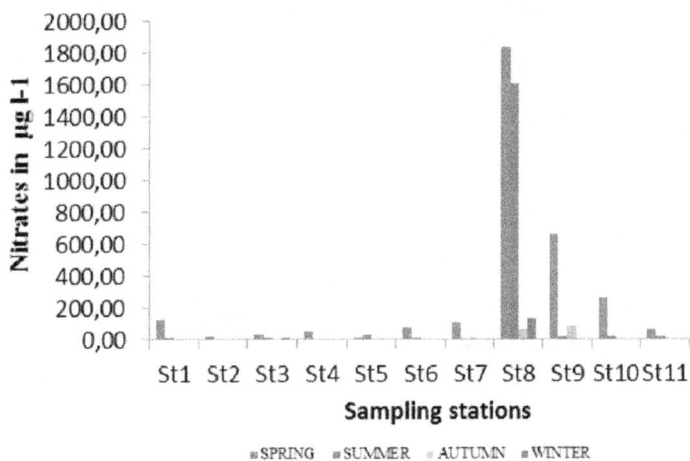

Figure 8.
Spatio-temporal evolution of nitrates at Dakhla Bay.

3.1.8 Phosphates

For the phosphate parameter, the values show a maximum of 0.37 mg l^{-1}, a minimum of 0.00 mg l^{-1} and an average of 0.14 mg l^{-1} (**Figure 9**). The results of previous studies have shown that phosphate levels in surface water have been in the range of 0.00 to 0.01 mg l^{-1} [25]. A clear trend toward increasing phosphate levels in the waters of the Bay of Dakhla has been noticed. This increase may be due, on the one hand, to the activities related to the processing industry of fishery products and domestic discards and, on the other hand, to the upwelling phenomenon that characterizes the Dakhla area [26].

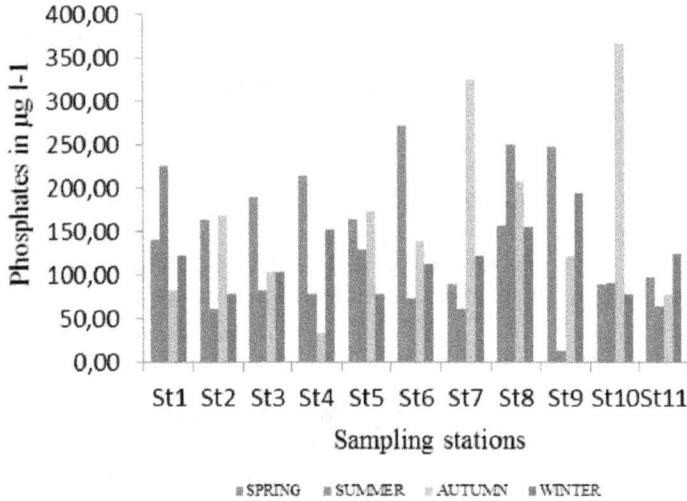

Figure 9.
Spatio-temporal evolution of phosphates at Dakhla Bay.

3.1.9 Chlorophyll (a)

Figure 10 shows the evolution of the chlorophyll (a) concentration during the study period, the mean value obtained was 2.74 µg l^{-1}, with a minimum of 0.00 µg l^{-1}

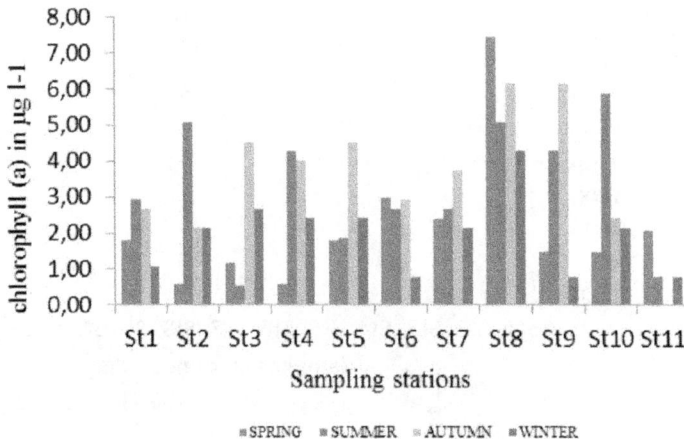

Figure 10.
Spati-temporal evolution of chlorophyll (a) at Dakhla Bay.

at "Puertitou" and a maximum value of 7.42 µg l^{-1} recorded in the urban area. These results indicate that the waters of Dakhla Bay have higher chlorophyll (a) levels than those reported in previous studies of the same ecosystem, which have showed that this concentration did not exceed 2.60 µg l^{-1} in 1991 and 5.00 µg l^{-1} in 1994 [25]. The high concentration recorded in this study corresponds to the site that is close to the urban area, which is undergoing essential nutrient enrichment for chlorophyll proliferation.

3.1.10 Water quality

Diffuse or punctual anthropogenic inputs have been responsible for significant nutrient enrichment (phosphates and nitrates in this case). These inputs come from various sources: agricultural, industrial or urban. Their "evacuation" or "elimination" is linked to the dilution capacity of the system, the hydrodynamics and the efficiency of the degradation processes of these elements by bacteria [24].

The implementation of the WFD has been a driving force forcing Member States to define environmental quality indicators associated with quality thresholds. These indicators concern dissolved oxygen, nutrients and phytoplankton, qualified among others through the chlorophyll concentration (a) [27]. Ifremer also defined in 2010 a quality indicator for coastal and transition masses, the nutrient indicator proposed by [10] integrates the dissolved inorganic nitrogen (NID) concentrations which include ammonium, nitrates and nitrites. Quality grids proposed by the European Water Framework Directive [9], enabled us to classify the waters of Dakhla bay as good quality for all the physico-chemical parameters studied, except for the sites located near urban discharges which have an average quality (**Table 1**).

REF	T	pH	O2	NID	PO4	Chlo
St1	Good	Good	Very good	Good	Good	Very good
St2	Good	Good	Very good	Good	Good	Very good
St3	Good	Good	Very good	Good	Good	Very good
St4	Good	Good	Very good	Good	Good	Very good
St5	Good	Good	Very good	Good	Good	Very good
St6	Good	Good	Very good	Good	Good	Very good
St7	Good	Good	Very good	Good	Good	Very good
St8	Good	Good	Very good	Average	Good	Good
St9	Good	Good	Very good	Average	Good	Very good
St10	Good	Good	Very good	Good	Good	Very good
St11	Good	Good	Very good	Good	Very good	Very good

Table 1.
Dakhla Bay Water Quality Grid

3.2 Sediment compartment

3.2.1 Grain size

In Dakhla Bay, Rudites are variable in the sediments. They range from 0.60% to 3.94% of the total dry sample weight. These Rudites mainly come from the biogenic fraction consisting of lamellibranch and gastropod shells [28]. The particle size distribution of the sediments collected at the Dakhla, shows an abundance of Arenites with percentages ranging from 75.60% to 98.28% respectively for the Port Basin and

Boutalha sites. This result informs about the hydrodynamic aspect of the bay and shows that the currents in the downstream part are stronger than in the upstream part. For the Lutites fraction, the sediments contain 1.56% for Boutalha and 22.01% for the harbour basin. The sediment in Dakhla Bay therefore has a predominantly sandy and sandy-muddy particle size texture (**Figure 11**).

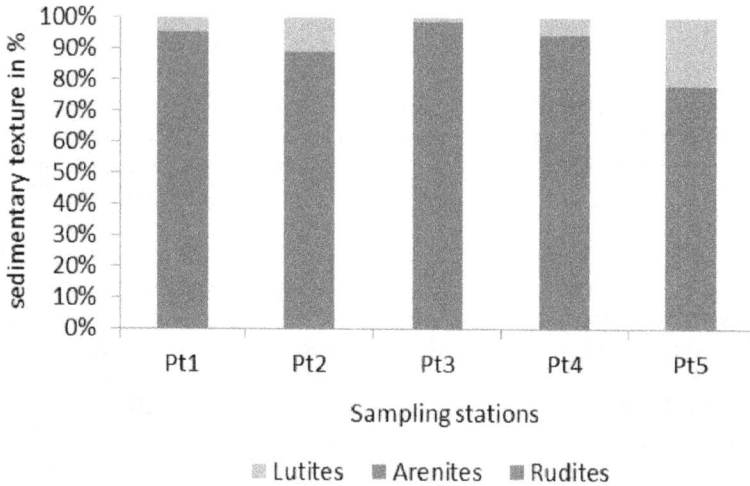

Figure 11.
Spatial evolution of the sedimentary texture of Dakhla Bay.

3.2.2 Organic carbon

During the present study, the percentages recorded for organic carbon oscillate around a minimum of 0.20% in Boutalha and a maximum of 2.85% in the port basin (**Figure 12**). Examination of the results of this study shows that the percentage of organic carbon gradually increases from the downstream to the upstream part. We find that the percentage of organic carbon follows the same evolution as the distribution of fine sediment fractions. The percentage of organic carbon in the harbour basin is ten times greater than in the bay, due to the confinement of the

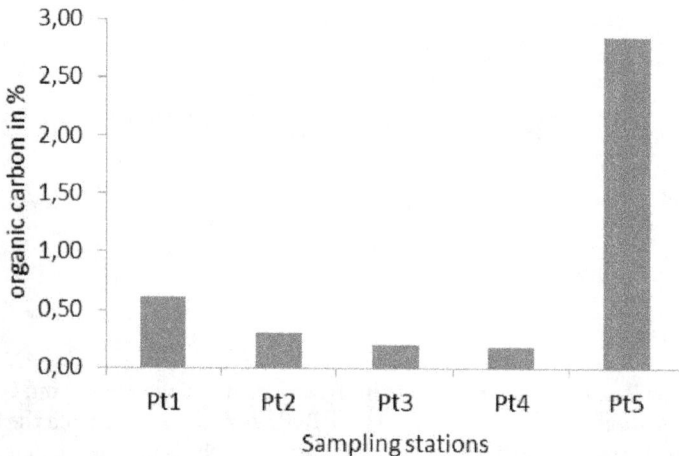

Figure 12.
Spatial distribution of organic carbon in the sediment of Dakhla Bay.

basin on the one hand, and on the other hand to the organic-laden discharges that could be dumped into the harbour.

3.2.3 Metal trace elements

The results for the majority of the metallic trace elements studied (cadmium, lead, chromium, copper and zinc) show a minimum at the Dunablanca station, except for mercury, for which a minimum was measured at the Boutalha station. On the other hand, the maximum was recorded, for all elements, at the Port Basin level. With regard to the three metals recognised as toxic (Cadmium, Lead, Mercury), the results recorded for Pb are of the order of 5.15 ± 0.12 and 16.69 ± 0.25 mg kg^{-1} successively in Dunablanca and Urban Area, while at the level of the port basin the concentration is 45.58 ± 0.61 mg kg^{-1}. For Cd concentrations are between 0.43 ± 0.01 and 0.62 ± 0.01 mg kg^{-1} respectively in Dunablanca and Boutalha, in the harbour basin the concentration is 1.3 ± 0.02 mg kg^{-1}. Mercury levels are below the detection limit for Boutalha and of the order of 0.0023 ± 0.0001 and 0.0145 ± 0.0021 mg kg^{-1} successively in the Urban Zone and the port basin (**Figure 13**).

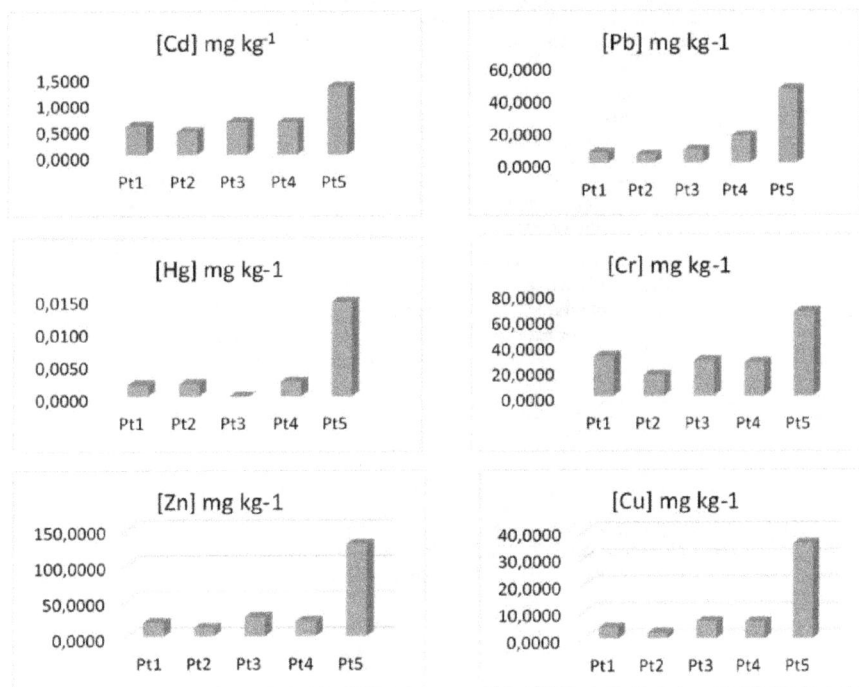

Figure 13.
Spatial evolution of trace metals in the sediment of Dakhla Bay. All values are in milligrams per kilogram (mg kg^{-1}) of dry sediment.

3.2.4 Sediment quality

Sediment quality in Dakhla Bay was assessed through the determination of particle size, total organic carbon content and the contents of the main metallic trace elements (cadmium, lead, mercury, chromium, copper and zinc).

The results of this study allowed us to assess the degree of contamination of the sediment in Dakhla Bay according to the criteria established by the Canadian

Council of Ministers of the Environment (**Table 2**). For the four stations Dunablanca, Pk25, Boutalha and the Urban Area, concentrations are below the CSE or even below the CER (**Class II**). On the other hand, for the Harbour Basin, the concentrations of Pb, Cd, Cu, Cr and Zn are between CSE and CEO (**Class III**), but for the Hg concentration they are still below CER (**Table 3**).

The study showed that, apart from the port basin, Dakhla bay remains less polluted, either compared to national or international ecosystems (**Table 4**). However, special attention must be paid to minimising, or even stopping, all kinds of pollution, in order to better protect the bay.

Sediment Quality Criteria	Pb	Cd	Hg
ERC	18	0.32	0.051
CSE	30	0.67	0.13
CEO	54	2.1	0.29
CEP	110	4.2	0.70
CEF	180	7.2	1.4

Rare Effect Concentration (REC), a Threshold Effect Concentration (TEC), an Occasional Effect Concentration (OEC), a Probable Effect Concentration (PEC) and a Frequent Effect Concentration (FEC) according to the Canadian Council of Ministers of the Environment.

Table 2.
Marine Sediment Quality Criteria [13].

Criteria for quality	Class	Impact on the environment	Prevention of sediment contamination from discharges	Evaluation of the quality of the sites studied
<CEF ____ <CEP ____ = ou > CEO	Class III	Frequently observed biological effects	The probability of measuring adverse effects increases with the concentrations measured. Examine the problem: continue investigations to identify the source(s) of contamination and intervene if necessary on these sources in order to avoid an increase in contamination or a new inflow of contaminants.	- Port basin
< ou = CSE	Class II	Biological efects semetimes observed	The likelihood of sediment having an impact on the environment is low. Monitoring can be put in place to verify the evolution of the situation.	-Pk25 -Dunablanca -Boutalha - Urban area
< OU = CER	Class I	Rarely observed biological effects	Sediments are considered to have no impact. No action is required, except in cases where persistent, toxic and bioaccumulative substances (e.g., mercury) released into water bodies may accumulate in sediment and in the tissues of organisms	

Table 3.
Application of sediment quality criteria [13] of the sites studied.

Studies/Criteria	Pb (mg kg^{-1})	Cd (mg kg^{-1})	Hg (mg kg^{-1})	Reference
Dakhla Bay (min - max)	(5.15–16.69)	(0.43–0.62)	(< LD - 0.0023)	This study
New Port (Dakhla)	45.58 ± 0.61	1.3 ± 0.02	0.0145 ± 0.002	
Dunablanca	3.6 ± 1.8	< LD	n/a	[28]
Nador Lagoon (min - max)	(3–416)	(0–6.2)	—	[29]
Moulay Bou Selham Lagoon	22.4 ± 7.5	0.94 ± 0.32	—	[30]
Sidi Moussa Lagoon	33.0 ± 5.1	3.67 ± 0.64	—	[30]
Oualidia Lagoon (max)	2.5	0.7	0.08	[31]
Ebrie Lagoon	(7–250)	—	(0.0–2.2)	[32]
Oludeniz Lagoon	7	—	—	[33]
Piratininga Lagoon	66 ± 20	—	—	[34]
Bay of Bothnia	79	0.94		[35]

Table 4.
Comparison of Pb, Cd and Hg contents in sediments of Dakhla Bay with levels in other paralic environments.

3.3 Statistical analysis of results

More than 83% of the variation of all parameters studied is expressed by the two factorial axes F1 and F2 of the principal component analysis (PCA), with 58% for the F1 axis and 25% for the F2 axis. The principal component analysis shows that the parameters (Nitrates and chlorophyll (a)) are significantly correlated to the first factorial axis (F1 axis) in the urban area while the parameters (Total organic carbon, phosphates and trace metal elements) are significantly correlated to the second factorial axis (F2 axis) in the port area.

The analysis enabled us to highlight that the areas containing urban agglomerations and the port are the most impacted (**Figure 14**).

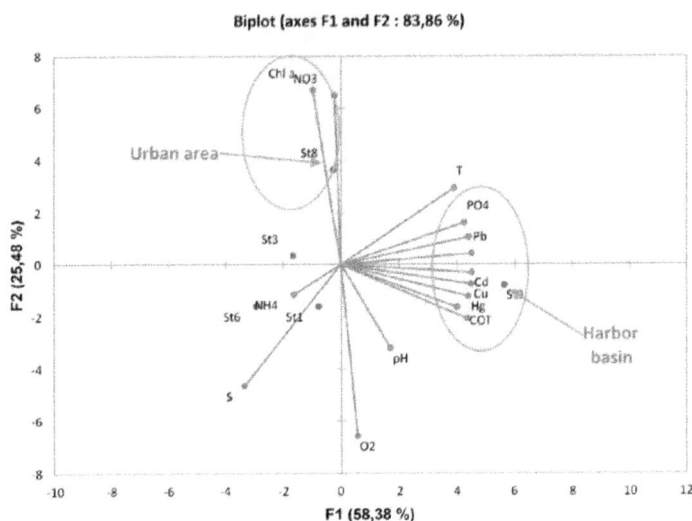

Figure 14.
PCA results for the variables studied for water and sediment of Dakhla Bay.

4. Conclusion

This study shows that:

The monitoring of physico-chemical parameters revealed nutrient enrichment and a significant chlorophyll biomass, especially at sites close to wastewater discharges.

Based on the different quality grids proposed by the WFD, a qualification of the waters of Dakhla Bay was established, which is generally "Good", except for two stations located near urban discharges, for which the quality is "Average".

The granulometric study enabled us to identify the sedimentary structure of Dakhla Bay, which is of sandy to sandy-muddy type.

The evaluation of the levels of metallic trace elements (Cd, Pb, Hg, Cu, Cr and Zn) in the sediments shows that, apart from the port basin, Dakhla Bay is less polluted, both in comparison with national and international ecosystems.

In conclusion, Dakhla Bay remains a site little impacted by human activities. However, particular attention must be paid to minimising, or even stopping, all kinds of pollution, to better protect the bay.

Author details

Mimouna Anhichem* and Samir Benbrahim
National Institute for Fisheries Research (INRH), Casablanca, Morocco

*Address all correspondence to: anhichemm@gmail.com

IntechOpen

References

[1] CHABAUD, C. (2013) - What means and governance for sustainable ocean management? The opinion of the economic, social and environmental council Official Journals. http://www.lecese.fr/sites/default/files/pdf/Avis/2013/2013_15gouvernanceoceans.pdf.

[2] Kouassi, Aka Marcel. (2005) - Hydrochimie and water quality in two tropical lagoons of Cote d'Ivoire (Ebrie, Grand Lahou).≫ These from Doctorate, University of Cocody,. Abidjan, 147 p.

[3] Ifremer (2008) DCE-MEFM - Taking into account the impact of shellfish farming activities on the biological quality element "Benthic invertebrate fauna" in the assessment of ecological potential - Case of the MEC "FRFC02". and " FRGC01 " Final Report.

[4] Maanan M. Sedimentological study of the filling of the sidi moussa lagoon (Moroccan Atlantic coast) granulometric, mineralogical and geochemical characterization. PhD thesis, Chouaib Doukkali University, Eljadida Faculty of. Sciences. 2003;**131p**

[5] Villanueva, Maria Concepcion S. (2004) - Biodiversity and trophic relations in some estuarine and lagoon environments in West Africa: adaptations to environmental pressures, PhD thesis from the Institut National Polytechnique de Toulouse, 246 p.

[6] Aminot A. and Chaussepied M. (1983) - Manual of chemical analyses in the marine environment. Centre National pour l'Exploitation des Océans, 395 pp.

[7] Aminot A., Keroul R. (2004) - [Hydrology of Marine Ecosystems: Parameters and Analyses]. IFREMER edition, 336 pp.

[8] Taverny, C. & Pierre Elie. (2009) - [Review of biological knowledge and the state of lamprey habitats migratory migrants in the Gironde basin]. Proposals for priority actions. irstea., pp.110.

[9] WFD(2000/60 / EC). (2000) - The Water Framework Directive or European Directive of the European Parliament and of the Council adopted on October 23, 2000. Ifremer.

[10] DANIEL, A. & SOUDANT, D. (2010) - WFD Assessment April 2010 - Quality Element: Oxygen Balance. Ifremer Report, 73 p.

[11] Loire Bretagne Water Agency. (2005) - Qualification of coastal and transitional water bodies on the basis of environmental quality data. Result of the assessment. Progress report 4.

[12] SEQ-Water, Biology Aptitude Class (2003).

[13] BELIAEFF, B., BOUVET, G., FERNANDEZ, J.-M., DAVID, C. & LAUGIER, T. (2011) - Guide for the monitoring the quality of the marine environment in New Caledonia. ZONECO and programme CNRT the Nickel. 169 pages.

[14] CMEC (2007) - Environment Canada and the Department of Sustainable Development, Environment and des Parcs du Québec. Quebec Sediment Quality Assessment Criteria and Frameworks application: prevention, dredging and restoration, 39 p.

[15] RAMSAR. (2016).- The List of Wetlands of International Importance Published.

[16] QUINBA, A., RADI, M., BENHOUSSA, A., BAZAIRI, H. & MENIOUI, M. (2003) - Fact sheet on Ramsar Wetlands (RIS). Categories endorsed in Recommendation

4-7 (1990) as amended by Resolution VIII.13 of the Conference of the Contracting Parties, 9 p.

[17] BELLAIR, P. & POMEROL, C. (1977) - Elements of geology. Paris, Armand Colin.

[18] WALKLEY, A. & BLACK, I.-A, (1934).- An examination of the Degtjareff method for determining organic carbon in soils: Effect of variations in digestion conditions and of inorganic soil constituents. Soil Sci, 63, 251-263.

[19] OSPAR/JAMP, 2002.- Guidelines for Monitoring Contaminants in Sediment. Ref. No. 2002-16. htpp://www.ospar.org/documents?d=32743.

[20] CTAQUA-CETECIMA (2014) - Report on the identification of the characteristics of the physical environment. Plan for the planning and development of aquaculture in the Oued Eddahab region Lagouira - Morocco Lot 1 Phase 2.

[21] ORBI, A., GUÉLORGET, O. & LEFÈBVRE, A. (1996) - La baie de Dakhla, organisation biologique et operation. INRH, Dakhla, Morocco, 240 p.

[22] DAFIR, J. (1997) - Application of phosphorus dynamics to the study of organization and functioning. aquatic ecosystems (management and preservation). PhD Thesis, University Hassan II, Faculty of Science Ain Chock, Casablanca, 634 p.

[23] SAAD, Z. (2015) - Assessing the quality of coastal marine waters: a new methodological approach applied for the case of Dakhla Bay (South Morocco). [Thesis] Ecology and management Cadi Ayyad University, Faculty of Sciences, Semlalia, Marrakech.

[24] SOUCHU, P., LAUGIER, T., DUSSERRE, K. & MAROBIN, D. (2001) - Monitoring of trophic parameters. in the water of the Narbonnaise ponds. Final report, Ifremer DEL, Coastal Laboratory of Sete, February 2001.

[25] ORBI, A., DAFIR, J.M. & BERRAHO, A. (1995) - Multidisciplinary study of Dakhla Bay. Works and document n°86 from INRH, Morocco. Internal report, 26 p.

[26] MAKAOUI, A., ORBI, A., HILMI, K., ZIZAH, S., LARISSI, J. & TALBI, M. (2005) - L'upwelling de la Atlantic coast of Morocco between 1994 and 1998. C. R. Geoscience, 337, 1518-1524.

[27] FOUSSARD, V. & ETCHEBER, H., (2011) - Report. Proposal for a monitoring strategy for physico-chemical parameters for the Seine and Loire estuaries. BEEST project: towards a multi-criteria approach to the good ecological status of large estuaries. Univ. Bordeaux 1, UMR 5805 EPOC and CR1 CNRS.

[28] ZIDANE, H., ORBI, A., MOURADI, A., ZIDANE, F. & BLAIS, J.-F. (2008).- Hydrological structure and edaphic of an oyster-farming site: Dunablanca (The bay of Dakhla in southern Morocco). Environmental Technology, 29, 1031-1042.

[29] RUIZ, F., ABAD, M., OLIAS, M., GALAN, E., GONZALEZ, I., AGUILA, E., HAMOUMI, N., PULIDO, I. & CANTANO, M. (2006).- The present environmental scenario of the Nador lagoon (Morocco). Approximately. Res., 102, 215-229.

[30] CHEGGOUR, M., CHAFIK, A., LANGSTON, W.J., BURT, G.R., BENBRAHIM, S. & TEXIER, H. (2001).- Metals in sediments and the edible cockle *Cerastoderma edule* from two Moroccan Atlantic lagoons: Moulay Bou Selham and Sidi Moussa. Approximately. Pollut, 115, 149-160.

[31] INRH (2015) - Lagoon of Oualidia. Ecological status and environmental

health. Integrated rehabilitation programme for the lagoon of Oualidia. INRH, Casablanca, Morocco.

[32] KOUADIO, I. & TREFRY, H.H. (1987).- Sediment trace metal contamination in the Ivory Coast, West Africa. Water Air Soil Pollut. 32, 145-154.

[33] TUNCEL, S.G., TUGRUL, S. & TOPAL, T. (2007).- A case study on trace metals in surface sediments and dissolved inorganic nutrients in surface water of Ölüdeniz lagoon -Mediterranean, Turkey. Water Res., 41, 365-372.

[34] LACERDA, L.D., FERNANDEZ, M.A., CALAZANS, C.F. & TANIZAKI, K.F. (1992) – Bioavailability of heavy metals in sediments of two coastal lagoons in Rio de Janeiro, Brazil. Hydrobiologia, 228, 65-70.

[35] LEIVUORI, M. (1998).- Heavy metal contamination in surface sediments in the Gulf of Finland and comparison with the Gulf of Bothnia. Chemosphere, 36, 43-59.

www.ingramcontent.com/pod-product-compliance
Lightning Source LLC
Chambersburg PA
CBHW081543190326
41458CB00015B/5623